How to Keep

A Handbook for of Beginners

By
Anna Botsford Comstock, B. S.

With Illustrations

PREFACE

THIS book has been prepared especially to meet the needs of the beginner in bee-keeping. It is not intended to be a complete treatise for the professional apiarist, but rather a handbook for those who would keep bees for happiness and honey, and incidentally for money. It is hoped, too, that it will serve as an introduction to the more extended manuals already in the field.

When we began bee-keeping we found the wide range of information and varying methods given in the manuals confusing; but a little experience taught us that bee-keeping is a simple and delightful business which can be carried on in a modest way without a great amount of special training. After a beginning has been made, skill in managing the bees is gained naturally and inevitably, and interest is then stimulated by the wider outlook which bewilders the novice.

For the sake of simplicity this volume is restricted to knowledge gained in practical experience in a small apiary; and the writer has sought to exclude from it those discussions which, however enlightening to the experienced, are after all but devious digressions from the simple and straight path which the feet of the inexperienced must tread to success in the apiary.

Photograph by Ralph W. Curtis
PLATE I. BASSWOOD BLOSSOM

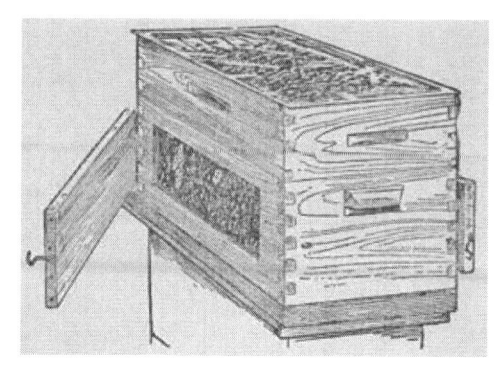

An ordinary beehive made into an observation-hive by inserting glass panes in sides and putting a glass sheet under the wooden cover. (From V. L. Kellogg.)

An observation-hive holding only two frames, with the two sides wholly of glass, so that any single bee can be continuously watched. (Form V. L. Kellogg.)

CHAPTER I
WHY KEEP BEES

THE reasons for keeping bees are many and various; for it is an industry as many-sided as the cells of a honey-comb, one of its chief charms being that it appeals equally to "many men of many minds." To know all the reasons why one should keep bees one must be conversant with the history of man; he must be familiar with the Vedas of India and master of the hieroglyphics of ancient Egypt; he must study the life of the common people in the Hebrew Scriptures and the classic literatures of Greece and Italy, and must be able to translate into terms of common human experiences the myths and legends of diverse peoples. For bees have been a part of the conscious life of man from the beginning. Not only have they sweetened his daily bread with beneficent honey, but they have had their place in his religious rites, and also in his family observances, as millions of happy lovers may attest who have experienced the "honey-moon." However, lacking the erudition necessary to enumerate all, the author of this book is quite content to mention a few of the more cogent reasons why anyone should, in this day and generation, undertake the keeping of bees

One may keep bees for the sake of the honey, which is a most legitimate and fit reason. Honey is a wholesome and delicious addition to the family table. Though it is a luxury, yet it may be afforded by one living in moderate circumstances, if he be willing to give a modicum of time and care to the happy little creatures that gladly make it for him, free of all expense. But for the one possessed of great wealth, and able to corner the honey-market any day, there is sure delight, as well as education, in raising his own honey. Any person of experience well knows that the honey made by one's own bees has quality and flavour superior to that made by other people's bees. In fact, the only way to become a connoisseur of honey is to keep bees, for thus only may one learn to discriminate between honey made from basswood and that gathered from clover; or to distinguish, at first taste, the product of orchard in bloom from that drawn from vagrant blossoms, which, changing day by day, mark the season's processional. He who has once kept bees and become an artist in honey-flavours, would never again, willingly, become a part of the world at large, which, in its dense ignorance, dubs all white honey "basswood" or "clover" indiscriminately, and believes all dark honey is gathered from buckwheat.

Some, perhaps, might keep bees for the sake of making money. For him who would get rich keeping bees, this book manifestly is not written; its title explains that it is meant to show how to keep bees, and not how to make bees keep him. The person who would, from the first, make bee-keeping his chief work should receive his training in a large apiary. As a vocation it demands the entire time and energy of a shrewd and able person to insure success; of such, America already has a great number, with yearly incomes varying from $1,000 to $10,000. However, the desire to make the bees keep themselves and add more or less to the family income is a practical and sensible reason for keeping bees. Fifteen or twenty colonies may be managed with comparatively little time and attention and the work may be done largely by women or the younger members of the household. If proper care be given to such an apiary, it will prove of material benefit to the family purse; for, if the season be favourable, the product of one colony should net the owner from four to ten dollars. We know of boys who have thus earned their college expenses; and many women have bought immunity from the drudgery of the kitchen with money paid them for their crops of honey. It should be borne in mind that honey-money is not obtained without thought, energy, and some hard work. The bees would have been less beneficent to mankind had they bestowed honey without demanding a return in care and labour.

Many have kept bees as a recreation, and there is none better. It gives delightful and absorbing occupation in the open air and is not merely a rest from mental and sedentary labours, but is a stimulus to health and strength as well. In the various bee journals are recorded testimonials from thousands who, when tired, ill, and nervously worn out, began bee-keeping, and through it regained vigour and found a new interest in life. Notwithstanding the fact that bees have honest stings, they can work the miracle of changing a dyspeptic pessimist into a cheerful optimist with a rapidity and completeness that merits our highest admiration.

A love for natural science is a very good reason for keeping bees. Nowhere are there more fascinating problems for the investigator than those still unsolved in the hive. Of all the lower animals, bees are, perhaps, the most highly developed in certain ways; and it is more probable that the mysteries of the eternal heavens and the distant stars be made plain than that we ever learn the truths that underlie the development of the "Spirit of the Hive."

Another reason for keeping bees is the insight to be gained therefrom into the conditions of perfect communism. The bees and their relatives are the most intelligent and consistent socialists that have yet been developed in this world; and, through studying their ways, one may discern with startling clearness how the perfect socialism grinds off all the projecting corners of the individual until it fits perfectly in its communal niche. In the hive individual traits, as exemplified in kindness, selfishness, love and hate, are moved up a notch in the scale and characterise the whole community, even though they are eliminated from its members. If one is a social philosopher, he may become very wise, indeed, by studying the results of the laws of socialism which have been executed inexorably through countless centuries in the bee commune.

Last, but by no means least, one may keep bees because they belong to home life and should have place in every well-kept garden or orchard. There is no more beautiful domestic picture to be found in the world than a fine garden with a row of hollyhocks hiding the boundary fence and affording a fitting background for a neat row of white hives. It is not alone the æsthetic beauty the bees bring to the garden which touches the deeper currents of feeling and is productive of profound satisfaction; it is something more fundamental; for since the thyme of Hymettus yielded nectar, the happy hum of the bee has yielded comfort to the human senses. The garden without bees seems ever to lack something; mayhap, the silent longing of the flowers for their friendly visitors, intangible but real, so permeates the place that we are dimly conscious of it. Be that as it may, the perfect garden can only be attained through the presence of happy and populous bee-hives.

CHAPTER II

HOW TO BEGIN BEE-KEEPING

THE VARIOUS WAYS OF BEGINNING

THERE are so many ways of beginning this interesting work that no classic way obtains. Many people have received the stimulus from a swarm of bees, escaped from some apiary, which has alighted on a tree or bush on the premises, and which seemed too much like a gift of the gods to be ignored. In fact, no one with blood in him would do otherwise under such circumstances than to hunt up a soap-box or a nailkeg and, with an intrepidity amounting almost to heroism, place it under the cluster and shake the bees into it. Then, if the swarm feels content, the fact is accomplished; and the involuntary owner finds a new interest in life, and enthusiastically becomes a bee-keeper. This is an excellent and inexpensive way to begin, and for all who are thus favoured it is by far the best way. But one may wait many years before this happens, and after all there are other and more direct methods. The best way is to begin by wishing sincerely to keep bees, and then to adopt one of the following plans:

The cheapest way is to visit the nearest neighbour who keeps bees, and buy of him a swarm, which will cost, perhaps, two or three dollars. If the neighbour be a good apiarist, this may be an excellent method, for he will give practical advice and be a most helpful friend in time of future difficulties and doubts; he will also explain appliances and make the labours and perplexities of the beginner much smoother. Besides these advantages, it is a help to neighbourliness, for keeping bees is almost as close a bond between two neighbours as an interest in golf or automobiles, and has a much broader and more philosophical basis.

Of course, the bees may be bought at any time of the year convenient, but the early spring is the best season for beginning, for then one has the advantage of increase by swarms. If one is ingenious and inventive, one may easily construct other hives like the one bought; but there are other things needed which may be obtained cheaper and better from firms that sell apiarists' supplies. However, we are acquainted with several apiarists who furnish all of their own supplies except the sections for comb-honey, the wax foundations, the smoker, the cartons and the honey-extractor.

The usual way of beginning is to send to some of the dealers in bee-keepers' supplies for a catalogue and to invest in a library of bee books. There is something most fascinating about these books written by bee-keepers, for no one makes a success of bee-keeping unless he loves the bees; and if he loves bees he writes of them so persuadingly, and his lines are so full of insidious allusions to the enchantments of the occupation, that whoever reads one of these volumes finds life arid without bees.

Then the catalogue of bee-keepers' supplies has its own attractions. Never were such breathlessly interesting pamphlets written; and I would add that, on the whole, they are the most honest and reliable of all advertising catalogues. They almost invariably give good and sensible advice to beginners, suggesting proper outfits at moderate prices. To the would-be bee-keeper these catalogues are so delightful that, if the purse is long enough, he feels inclined to order a specimen of everything listed. However, this is not the best way to do. A moderate number of things can be ordered at first, and other things may be ordered later as the growing sense of their need develops.

If one is minded to go into bee-keeping as a regular business, the best way to begin is to spend a year with a successful bee-keeper, working for him and with him and paying for the tuition, whatever may be charged. Thus one may gain his knowledge of the business and get his practical training under the guidance of experience. The very worst way to begin, and one that is sure to result in disaster, is to buy a large number of colonies at first. This is a too expensive way of learning the extent of one's own ignorance and limitations as a bee-keeper.

As we have begun keeping bees three times in the course of our lives, we feel more or less competent to give advice about this special phase of bee-keeping. We first began when we were children and two swarms were bestowed upon us by an uncle who moved away from the neighbourhood. This was long before the idea of supers and box comb-honey had been thought of. We were among the first to adopt the old glass-sided super when it was introduced. The second and third times we purchased our outfit of a dealer in apiarists' supplies, and both of these beginnings were good and successful; so perhaps the best way to tell others how to begin is to describe how we began. The following is the order which we sent to a dealer.

A BEGINNER'S ORDER

One colony of bees, in an eight-frame, dovetailed, chaff hive with a deep telescope cover, and with a tested Italian queen.

- Two complete supers ready for use, with 4¼ x 4¼ x 1½ plain sections.
- One super cover.
- Three extra 1½-story chaff hives uniform with the above, and with super covers.
- One standard Corneil smoker.
- One bee-veil.
- One pair bee-gloves.
- Two hundred and fifty sections, plain, no bee-way, 4¼ X 4¼ X 1½.
- One pound medium brood foundation.
- Two pounds thin super foundation.
- One Daisy foundation-fasteners
- One Van Deusen wax-tube fastener.
- One ¼-pound spool No. 30, tinned wire.
- One Porter bee-escape, with board.
- One Manum's swarm-catcher.
- One Dixie bee-brush.
- One Doolittle division-board feeder.
- One Alley's queen and drone trap.

NOTES ON A BEGINNER'S ORDER

THE BEES

The selection of the bees is the first and most important consideration, since race and heredity determine to so great an extent the bee's efficiency and disposition.

The consensus of opinion to-day is in favour of the Italian bees; this is so much so that the other races, except the German, are hardly on the general market in America. Taking all points into consideration, the Italian has a higher average of satisfactory qualities than has any of the other races.

Our earliest experiences were with the ordinary black German bees, and it was through them that we first learned to love bees, although their nervousness and their unhappy habit of considering us intruders when we neared their domain were always somewhat embarrassing, and made us feel like James Whitcomb Riley's visiting base-ball team, that "we weren't so welcome as we aimed to be." Therefore, when we decided to buy bees, we un-hesitatingly ordered Italians. These are much more gentle and friendly than the others, and it is one of our greatest pleasures to be on good terms with our bee-folk. Under all ordinary circumstances the Italian bees are not only tolerant of human interference, but are sweet tempered and optimistic, believing that we mean well by them; and one cannot work with them without learning to love them.

If one begins bee-keeping in the spring, some money can be saved by buying a nucleus and tested queen, instead of a full colony of bees. A nucleus in bee-keeping parlance is a small colony with only one, two, or three frames. The colony should be large enough to build combs with sufficient rapidity to keep the queen fairly busy, so as not to encourage in her the habit of loafing, and so that too much time shall not be required to build up a full colony. Two frames with about a quart of bees will accomplish this well.

Such a nucleus as this in a hive, with a division board on either side of the two adjacent frames, can be soon built up into a full colony if there is plenty of nectar and pollen to be had. A division board is a solid board of the shape of the frame, but a little larger so that it is close-fitting in the hive. The frame should be watched closely, and as soon as the comb is filled with brood, and there are enough bees to cover the brood well, another frame filled with brood foundation should be added; this should be continued until the hive is filled with frames.

A TESTED QUEEN

As the queen is the mother of the whole colony, her powers for transmitting a kind disposition and superior honey-making qualities are of the greatest importance. It is essential, therefore, to know that the queen, which is to serve as a foundation for the apiary, has mated with a drone of the desired race. A tested queen is one that has been kept until some of her offspring have been reared. The colour of these will indicate whether the queen has mated with a drone of her own race.

THE HIVE

Next to choosing the bees comes the selection of the hive, for there are several kinds of hives in general use, and all of them are apparently good. We chose the Langstroth hive because its merits are attested by the fact that it is more largely used than all others taken together. We chose the eight-frame in preference to the ten-frame form because we wished to produce comb-honey, and it is easier to induce bees to store surplus in the sections when the brood-chamber is small. If one wishes to produce extracted honey, the ten-frame hive is the better one. We ordered the more expensive chaff hives, as we wished to save the trouble of moving the bees into winter quarters; and we think this would be the case with anyone with whom bee-keeping is an avocation instead of a vocation. In fact many

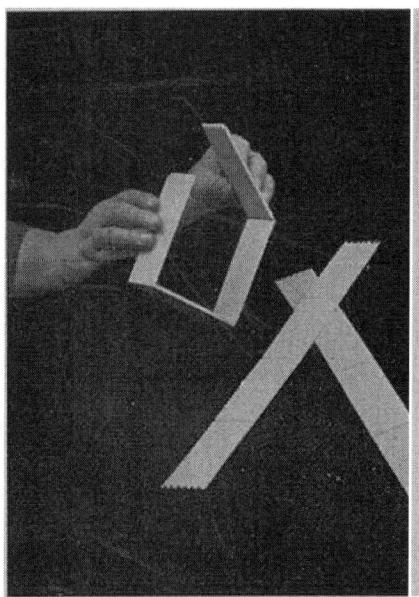

(*a*) Two flat section-boxes; and one being bent together.

(*b*) A Corneil smoker.

PLATE II

(a) The Dixie bee-brush, spur wire-embedder, and Van Deusen wax-tube fastener.

(b) A super with fences and two rows of section-boxes in place; a separate fence.

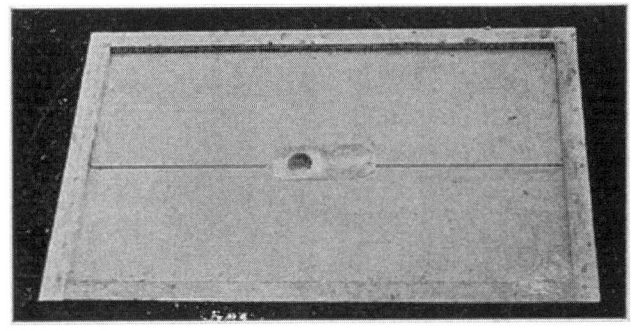

PLATE III.(c) The Porter bee-escape in a honey-board.

people who make bee-keeping their principal business prefer chaff hives, for they not only keep out the cold in winter, but also protect the bees from the heat in summer. The objections to them are that they are heavy to handle and are expensive, though the expense can be reduced considerably if one has the ability to make them. Another thing we like about the chaff hives is that they are fine and dignified in appearance, and we find that their majestic proportions, white and beautiful, set against the background of our larches, add much to our conscious pleasure every time we look upon our apiary. We ordered a deep telescope cover, as we wished room for two supers on the hive at once.

The hive-stand we ourselves made from lumber taken from dry-goods boxes. It is simply a smooth platform, six inches wider than the hive on three sides and extending about a foot out in front of the hive, thus serving as an alighting-board.

TWO COMPLETE SUPERS

We knew that we would need these very soon, for the basswood harvest was imminent; we ordered them ready for use, as we wished to see all the "new wrinkles" in supers, and exactly how the parts are arranged. We used the supers as models later in putting together and fitting other supers. This proved a wise precaution, as it saved us much time in reading directions and studying out independently the arrangement of parts. We ordered $4\frac{1}{4}$ x $4\frac{1}{4}$ x $1\frac{1}{2}$ sections, as this is the size most generally used, and we chose the no-beeway, or "plain" style, because we think it presents a better appearance when filled with honey. We will return to this question in Chapter VIII.

THREE EXTRA HIVES

We wished to have these on hand for any swarms which might possibly issue.

ONE STANDARD CORNEIL SMOKER

I do not know just why we chose this of the many excellent smokers, but perchance the name called to mind happy experiences with sundry Cornell smokers of quite a different feather, and we were thus favourably disposed toward this one. It has not disappointed us in the least, for it is both handy and practical. It may seem an unfriendly act to smoke one's own bee-people, but a little smoke wisely applied is as efficacious in preserving pleasant relations with the bees, as was the smoke from the pipe of peace in preserving similar relations between our forefathers and the savages. (Plate I.)

ONE BEE-VEIL

The senior partner of our apiary rarely uses a veil, but when he does use one he needs it very much, and it is an article necessary to have at hand. To the beginner it gives a calmness of nerve and a surplus of courage which are highly desirable when dealing with such high-strung creatures as bees. Also there always occur times during the year when the bee-tempers are on edge for some reason or other; and at such times, if one be intrenched behind a bee-veil, it facilitates work and encourages a serene spirit.

ONE PAIR OF BEE-GLOVES

While we do not use these ordinarily, yet when we have some special work to do which involves changing many bees from one location to another, we find these gloves most convenient to keep the disturbed little citizens from crawling up our sleeves, thus saving both them and ourselves from a most embarrassing situation.

250 SECTIONS

Only a small supply of sections was ordered, as but few would be needed the first year in addition to those in the two complete supers. (Plate II.)

ONE POUND BROOD-FOUNDATION

This was for use in the frames in the extra hives.

TWO POUNDS THIN SUPER-FOUNDATION

This was for use in the sections.

ONE DAISY FOUNDATION-FASTENER

In the early days of our bee-keeping we fastened the foundation into the sections and frames with a common kitchen-knife which we heated over a lamp and then applied to the edge of the wax foundation held against the section, thus melting it and pressing it fast to the wood. Afterward we used a Parker fastener, and found it a great improvement over the primitive method. But this Daisy foundation-fastener as described in the catalogue appealed to the modern spirit in us. When we tried to use the machine, we were bitterly disappointed at the end of five minutes, but that was because the iron was not hot enough to properly melt the wax. After a little we learned to hold the foundation on the plate just long enough to melt it to a proper consistency so that it adhered to the section as soon as it was dropped upon it. Then it was that filling sections was placed on the list of sports. The rapidity with which we filled four dozen sections almost took our breath away.

ONE VAN DEUSEN WAX-TUBE FASTENER

This was ordered under the impression that it would be needed for fastening the foundation in the brood-frames, but when the hives came we found that a much better method of fastening the foundations had been devised. This is described in Chapter VIII. (Plate III.)

TINNED WIRE

This is used for strengthening the foundation in the brood-frames, as described later.

ONE PORTER BEE-ESCAPE

The Porter bee-escape is a simple and most useful device. It is set in a thin board just the size of the top of the hive; in the middle is a bit of tin which forms a round pit on the upper side. The bees descend into this pit, and, trying to get out, push apart two strips of tin set at angles to each other, fastened at the ends, which act like a valve, letting the bee out but not permitting her to push back. This is put between the super and the hive in order to free the super of bees before removing the honey. This escape is also used on the doors and windows of workshops or extracting-rooms or other places where bees get in and it is desirable to get them out. (Plates III., XVIII.)

THE MANUM SWARM-CATCHER

We bought this because we liked the idea of it, but as yet we have never had occasion to use it; however, we never look at its long handle without being filled with a mad desire to try it on a provoking swarm of bees clustered in the top of a cherry tree.

ONE DIXIE BEE-BRUSH

This is an exceedingly useful instrument for brushing bees from frames and from sections. (Plate III.)

ONE DOOLITTLE FEEDER

Bee-keepers of extended experience consider this the best and most satisfactory kind of feeder in use for small apiaries. (Plate XII.)

QUEEN AND DRONE TRAP

This is not a necessity; we bought it in order to try experiments in preventing swarming by its use, and also to have on hand in case an excess of drones should be developed in any of our colonies. (Plate XIV.)

CHAPTER III

THE LOCATION AND THE ARRANGEMENT OF THE APIARY

WHERE to put the hives is the first question, and this must be determined by two or three conditions necessary for the health and comfort of the bees. Hives should be placed where the sunshine may reach them in the morning up to eight or nine o'clock, and in the afternoon from three to four o'clock. An old orchard such as is kept because of picturesque beauty rather than for its crop of apples is an ideal place. The clean-culture orchards of the modern horticulturists are undoubtedly more efficient as producers of apples and money; but we are always grateful that there are still remaining many fine old orchards, on sod ground where the trees, more or less gnarled and twisted, are a joy to the artistic eye. Little wonder that such a place is the ideal spot for the apiary; if the hives are grouped four or five together beneath one tree, the requirements of shade will be met.

If there is no old orchard, what then? A young orchard will do, unless clean culture is practised; in the latter case horses and cultivator will not be permitted on the domain of the bee-people. If no orchard offers, then a trellis of vines extending east and west, eight feet high, may shade a few hives, and may be a thing of beauty in the garden as well. Grape vines, hops, Virginia creeper, or any other rapidly growing vines will do. To the one who loves his garden, there will be many ways suggested whereby the hives may be placed to compass both comfort for the bees and joy to the beholder. We started an apiary at the north of the lilac trees, and made it a part of the lawn.

If no such happy position for natural shade is to be found for the hives, then one must have recourse to artificial shade or double-walled hives. A very good method of shading, much in vogue among the farmers of our country, consists of a few boards placed awning-fashion above the row of hives. This is not an attractive solution to the problem, although perhaps it might be made so if this method were ever resorted to by anyone with a sense of beauty; but usually it is limited to a simple cover consisting of two or three boards nailed together, slanting a little toward the back of the hives to shed rain, supported by four posts, which hold it a foot or more above the row of hives.

In California instead of boards a thatched roof is made for this sort of protection, and is ample enough to allow an aisle for the apiarist between the rows of hives set back to back.

Many people use a single shade board, which consists of slats fastened together by cleats made large enough to project a foot beyond the hive on either side. This is placed directly on top of the hive, and has to be weighted down with stones, and is therefore awkward to handle when working with the bees. If the climate is hot, or in any case, a double-walled cover to the hive is most excellent, since it affords a chance for free circulation of air between the two boards which form it. However, these double covers do not obviate the need of shade, and natural shade is the most desirable sort.

In case the region is exposed to high winds, there should be a windbreak around the apiary. Mr. Root, who is one of the greatest of American bee-keepers, uses for this a row of hardy evergreens which grow together into a solid hedge. In case a windbreak, either natural or made, is impracticable, a board fence about eight or ten feet high, built on two sides of the apiary, usually the north and west sides, will be found to serve the purpose. This fence may be made the trellis for vines.

In the rear of the village garden is an excellent place for bees. A high board fence as a boundary, and perhaps a barn at the side will act as a windbreak, while the fruit trees yield a grateful shade. We know several such modest apiaries which are most attractive in appearance. There are those who live in cities or towns who are filled with the beekeeper's ambitions; and even they need not despair. There are on record accounts of several small apiaries kept on the housetops of the owners, who believe that roofs are for more than mere protection. Where bees kept thus get their honey is a bee secret, but undoubtedly every flower in the region yields them tribute. If bees be kept in town, they must be placed on a roof or else a high fence must intervene between the hives and the highway, so that the plane of bee flight shall be set above the heads of horses and drivers; for these brave little honeymakers have never been taught to turn to the right, and so they often dispute the way with teams and usually come off victorious; and this might make the bee-keeper unpopular in his community.

Another necessity in the apiary is that the grass in front of the entrance to the hives be kept mown; otherwise many a heavily laden bee will experience loss or injury among interfering grass blades. It is not practicable, even if one were heroic enough to try it, to run a lawn mower nearer than four or five inches from the hives, so many bee-keepers place salt or coal ashes on the grass within this area. Mr. Root goes so far as to advise the use of sheep as automatic lawn mowers in the apiary, as nothing else can cut grass so short as does the sheep. People say "as silly as a sheep," but that is a silly saying, for many people may learn something of value about the management of bees from the sheep, which, when attacked by them, thrusts its head philosophically into a bush where the bees cannot reach the tender parts, and trusts to its wool to protect it elsewhere. As a matter of experience, sheep kept in apiaries are rarely stung at all.

In our own apiary, where it was not practical to mow close to the hives, we followed two methods:

When we had many bees we placed a rough board over the ground in the immediate front of the hive; when we had only a few swarms, it was one of our joys to get on our knees on cold days, when only a few adventurous workers were going into the field, and with shears cut the grass close to the ground; and this period spent on our knees was not penance, but joy. However, it might well get to be penance in a large apiary.

Having found the place for the apiary, the next thought is of hive stands. Many bee-keepers use a hive that has a combination bottom board and hive stand; this has an inclined plane up which the loaded bees may climb if they strike the ground. This is a device which also saves the lives of many bees in cool weather, when they would scarcely be able, through numbness, to reach the entrance of the hive otherwise. However, there are other bee stands which hold two or three hives, which are very comfortable in height for the work of the bee-keeper. But it is always well to remember that the opening of a hive should be low down, as it is easier for the weary wings to let the honey-weighted bee down than to lift her up to the doorway. We use a simple platform, with blocks under the corners, so that there may be circulation of air beneath, and extending about a foot out in front of the hive, thus serving as an alighting board.

The arrangement of the hives in the apiary is a subject which will pay for thought. When beginning, this is easy enough, as they may be arranged almost any way, so long as shade and short grass are assured. After the apiary grows they may be arranged in several convenient ways; one is to have the lines of hives facing each other, thus making an alley for the bees; while there is a safe passageway for the man in the rear of the two rows.

When there are only a few hives, it is best to have the entrances face the south. In fact, the entrance should never face northward in a climate as cold as that of New York State. There is one thing to bear in mind in arranging an apiary; make the groups under the trees individual, so that the bees will have no tendency to become confused as to the location of their own homes. If two face west, then let two others face east, or perhaps a group of three face to the south, etc.

When it becomes necessary, for any reason, to change the location of a colony, a board should be set against the hive, in front of its entrance. The bees, meeting this obstruction as they emerge from the hive, will fly about the hive for some time, and thus mark the new location, to which they will return. If this precaution is not taken, many bees will fly from the hive, directly into the field, without noticing the change, and will then return to the old location and thus be lost.

A honey-house near the apiary is a great convenience. If this is not practicable, the next best arrangement is a honey-room in house, cellar, or shed. Such a room is a necessity even in a small apiary. This room should be well ventilated and screens should cover the windows, and a swinging automatic screen protect the door; bee-escapes should be placed in door and windows.

The room should contain workbench and tools; a table, chairs of varying height, an oil-stove, and boxes or cupboards in which all of the apiary supplies, implements, and machinery may be stored and kept ready for use.

Photograph by Brown Brothers

PLATE IV. Hives well shaded by a tree, but the grass ought to have been cut long ago.

Fig. 1

Fig. 2

Fig. 3

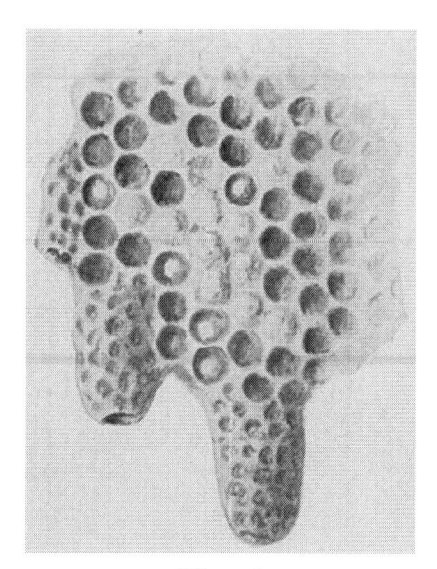

Fig. 4

Fig. 5

PLATE V. *1.* Queen Bee, enlarged. *2.* Drone. *3.* Worker. *4.* Queen Cells. *5.* Miller's Cage for Introducing Queens.

CHAPTER IV

THE INHABITANTS OF THE HIVE

ONE of the remarkable peculiarities of bees, which is also shared by other social insects, as ants, wasps and termites, is that there are three distinct kinds of individuals in the community. For, in addition to the males and females, which are the reproducing members of the colony, there is a third class which performs the labours of the community. The females, of which there is usually only one in a colony, are known as queens; the males as drones; and the labourers as workers. (Plate V.)

THE QUEEN (Plates V. VI, VII)

The bee-queen is the acme of a long line of communistic development. It is little wonder that those men of ancient times who observed her, and the attitude of the other bees toward her, regarded her as regal and called her queen. But she is a much more important element in the perfect commune than a mere sovereign, since she is the actual mother of her subjects.

Too much care cannot be shown in the selection of the queen, or mother of the bee-colony. Her blood is their blood; her faults are their faults; her weaknesses are their weaknesses. Any apiarist is likely to have had two colonies side by side, perhaps each equalling the other in amount of brood and number of bees, and one may have produced five dollars' worth of honey in a season, while the other did not produce half of that; and the queens alone caused this discrepancy. One produced energetic, capable offspring, while the progeny of the other were unenterprising. The offspring of one were perhaps sweet-tempered and obliging, and those of the other, cross and cranky. Thus it is all-important to give the colony a good mother. A queen, to be perfect, should be well-bred, handsome and strong, and capable of laying from two to three thousand eggs per day during the height of the season, and especially should she have offspring possessing a kindly disposition.

The laying queen is a very graceful insect; her body is long and pointed, and extends far behind the tips of her closed wings. *Svelte* is a graphic word applied to her figure by the *Spectator;* just a glance at her reveals her splendid physical development and proves her a queenlier bee than those that gather around her. It is a sight that makes men feel how very limited is their knowledge of any other world than their own to see the queen bee, surrounded by her ring of attendants, each with head toward her, as if she were the centre of a many-rayed star.

The development of the queen from the egg has ever been a most interesting and, at the same time, a most puzzling subject for investigation on the part of the scientists; even now after a century of study her growth is as miraculous as ever; and the problems in physiology that lie as yet unsolved in her development will keep many an investigator busy in the future.

When a colony is queenless, and has young brood or unhatched eggs, it makes haste to develop new queens; not one alone, but several, since it cannot afford to put its "eggs all in one basket." At the height of the honey season, every day that a colony is queenless means two or three thousand less bees than should be present to make it successful in securing the harvest. (Plate V.)

In developing a queen the bees usually proceed as follows: They select the important egg, which differs in no wise from any other worker egg, and destroying the partitions between its cell and two adjoining cells, give it more room. In order to make the royal apartment of good size a projection is built out over this large cell. This is made of thick wax and ornamented on the outside with hexagonal fretwork, as if it were to be the basis of comb with small cells. It seems as if the hexagonal pattern were in the bee brain and must be expressed, whether it be of any use or not. As soon as the little white larva hatches from the egg, it is fed on the regular larval food. Royal jelly is a food developed in the head glands of the workers; and when it is the fate of a bee larva to develop into a worker, it is fed with this food for three days, and then it is weaned by having other food substituted; but the queen larva is fed with it during her entire development, and thereby her reproductive organs are stimulated and fully developed, which is not the case with the workers. Think how much farther advanced are the bees than we, since, by giving the proper food, they are able to develop and fit each class of citizens to do the work required of it in the social organisation!

The queen larva is fed for five days on this most nourishing food, and then her cell is sealed. Within this cell the royal princess is for the first time self-dependent, and weaves about herself a silken cocoon and changes into a pupa. When she issues from this state she waits a little until she "finds herself," and then starts to cut an opening in the cell. She is a good mathematician, and with her jaws, cuts a circle very accurately, usually leaving it hinged like the lid to a pot. Professor Kellogg tells us that some-times when she cuts this door, the workers do not wish her to come out. They accomplish their purpose by carrying wax and pasting it over the opening as fast as she cuts it, at the same time quite devotedly feeding her through a small crevice. But if they wish her to come out, they rush to assist her, and perhaps for two or three days before she issues, make the wax thin where she is to cut. It usually requires sixteen days to develop a queen from the egg to the adult.

When a queen issues from her cell, she is light-coloured and, as her body is not yet distended with eggs, it is scarcely larger than that of one of the workers. Sometimes she chooses to stay in her cell for a day or two after it is opened. When she

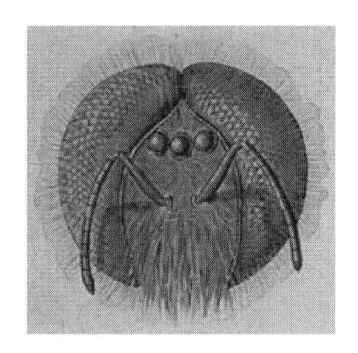

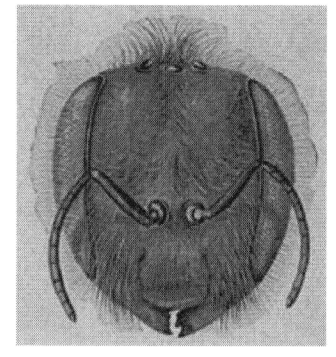

D   *Q*

W

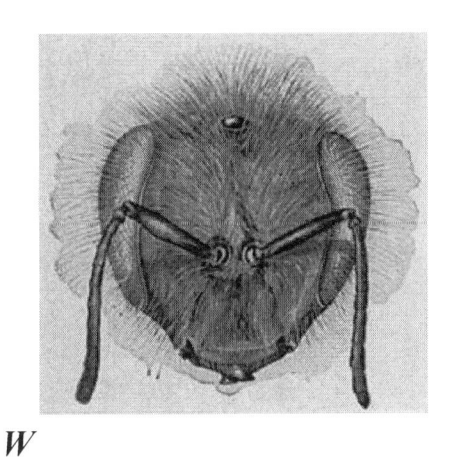

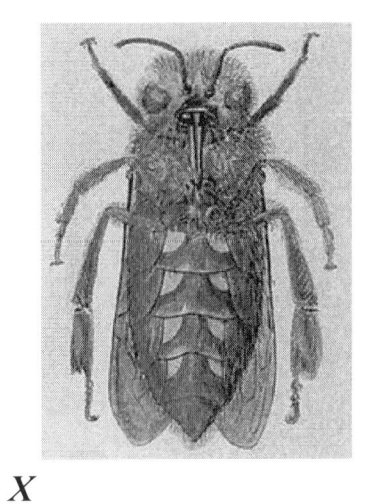

X

PLATE VI. (Original, drawn by A. G. Hammar.) *d*, head of drone; *q*, head of queen; *w*, head of worker; *x*, ventral surface of worker showing plates of wax.

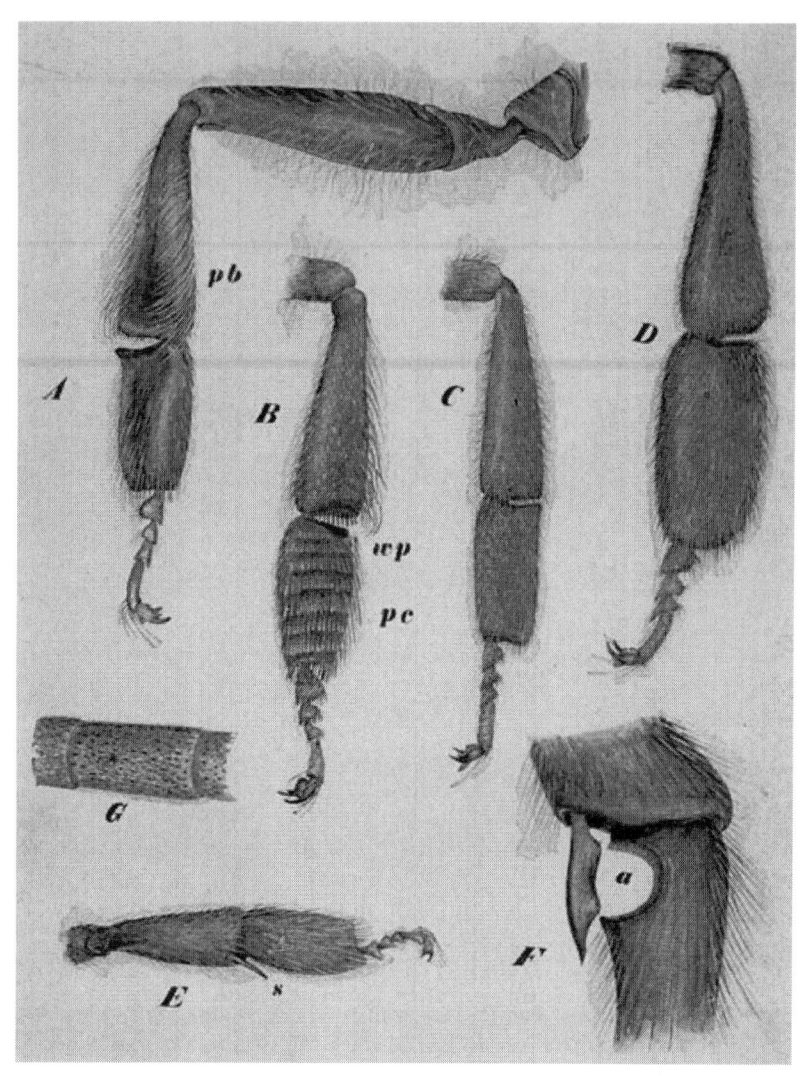

PLATE VII. Legs and antenna of the honey-bee (original drawn by A. G. Hammar). *A*, outer surface of hind leg showing the nine segments and claws; *pb*, the pollen basket of tibia; *B*, inner surface of part of hind leg; *wp*, wax-pincers; *pc*, pollen-combs; *C*, inner surface of part of hind leg of queen; *D*, inner surface of part of hind leg of drone; *E*, part of middle leg of worker; *s*, spur; *F*, part of fore leg showing the antenna cleaner *a; G*, part of antenna showing sense-hairs and sense-pits.

comes out, she runs about over the comb, taking exercise on her own royal legs, and perhaps taking a little honey out of the cells on her own account; especially does she hunt for other queen cells, for she has no wish to share her duties or honours with another. If she finds such a cell, she usually makes a hole in its side, and in some way, she stings to death the hapless princess within. Some observers claim that she merely takes the initiative, tearing down the wall of the cell, and the bees finish by tearing it down as they would any broken comb, and destroy the inmate in the process. If, in her promenade, she discovers another full-grown queen, a contest ensues; it is a duel to the death and the weapons are stings, which are kept sacred for this special occasion. It is interesting that the queen reserves her weapon for her peers, and never attempts to sting workers, and may be handled fearlessly by the bee-keeper. As the plate armour of the bee is so arranged that the sting may enter in only at certain spots, this duel resolves itself into a fencing match, until one thrusts her weapon into some vulnerable portion of the other. One morning we found fifteen dead queens outside of one of our hives; a grim tribute to the prowess of the queen within, and quite as much a tribute to our carelessness in letting so many queens be developed uselessly.

The belligerent attitude of the queens toward each other seems to have been so strong an emotion that a voice has been developed to express it, and is eloquent with rage and fear. This note must be heard to be understood; as nearly as I am able to spell it, it is "tse-ep, tse-e-e-ep, tse-e-ep, tsep, tsp, tsp, ts," in a sort of diminuendo. She makes the noise when she discovers another queen cell; if there is within this cell a full-fledged queen, she pipes back, but it sounds quite differently and the note is more like "quock, quock." This piping of queens is especially evident before an after-swarm is to issue. The queen will also pipe when the bees gather about her and try to ball her, which is often the fate of a new queen introduced into a colony not ready to receive her. In this case the note is one of righteous anger at the indignity to her royal person. She makes this piping with some vocal instrument, not well understood. Her wings vibrate tremulously while she is piping, but she can pipe quite as vociferously after her wings have been entirely cut off.

After she has made good her title to empire, the queen thinks about marriage; some warm day she will run out of the hive to see how the world looks, and especially to determine beyond doubt upon just what point of the universe her own hive is situated. The first flight of the queen bee is very pretty to see. She makes many graceful circles about, and plays in the sunshine as if she were thoroughly enjoying herself. When she finally leaves the hive to find a prince, she makes several little detours, always coming back so that she can commit to memory, beyond peradventure, the home place, and then off she goes in the sunshine to find her lover. Unfortunately she is not discriminating in the matter of love, and any sort of a prince, however lowly, is acceptable. Thus does many a fine, highly bred queen return to her hive, to bestow upon her progeny the undesirable traits of some low-bred drone. This is one reason why it is so difficult to keep an apiary of pure blood; and these *mésalliances* of queens are a source of much tribulation to the bee-keeper. She returns from her wedding journey with a part of the reproductive organs of her mate in her possession, often still visible, but soon after withdrawn into her body. With the sperm cells now under her control, she will fertilise the eggs of perhaps a million workers, more or less, which she may mother during her life of three or four years.

Biologists have of late achieved the miraculous in being able to stimulate the unfertilised eggs of sea-urchins and starfish, so that they will develop. The queen bee is able to do this with her own eggs. When the time comes for drones to be developed, she lays unfertilised eggs, which, unfailingly, produce the drones. If our poor human queens possessed this power of producing male heirs at will, much trouble would have been saved to many of them and, to some of them, their heads. However, the perfect socialists do most things better than we.

As soon as the queen returns from her honeymoon, which is usually taken from eight to ten days from the time she issues, she acts decidedly like a business person. She runs about on the comb, pokes her head into a cell to see if it is all ready, and then, turning about, thrusts her abdomen in and neatly glues an egg fast to the bottom. When the honey season is at its height, she works with great rapidity; sometimes she lays at the rate of six eggs per minute, often laying three thousand or twice her own weight of eggs per day. She is a wise queen, however, and has in mind the dangers of overpopulation. When there is much honey brought in, and the swarming season is at hand, she enlarges her empire rapidly; but when there is little honey, she keeps the amount of brood down to what can be cared for. Whether this question of limiting the population is decided by the queen, or whether she simply acts in response to the food given her by the workers, is a question not yet settled. However this may be, it is certain that Malthusian doctrines are rigorously and successfully practised by the perfect socialists of the hive.

Sometimes when the honey flow is very great an intoxication of work seems to possess the colony. The bees, coming in from the field, will drop the honey anywhere, and the queen, agitated by the general spirit of the hive, will drop her eggs everywhere; and the poor, overworked, bee housekeepers have to pick up the honey and store it in the cells, and pick up the eggs and glue them fast to the cell-bottoms.

THE DRONE (Plates V, VI, VII)

Of all the denizens of the hive the lot of the drones is the least enviable. That one may surely fulfil the destiny as king father, many are born, only to be slain when the honey harvest runs low, and meanwhile they are denied all interests in the business of the colony except that of pensioners upon its bounty. And he, the fortunate one, who lives his life to its fullest measure and becomes queen consort must in the end lose his life for love and die, heartlessly abandoned by her whom he sought and won.

In appearance the drone differs much from the queen and the worker; he is broad, and the rear end of his body is so blunt that it looks almost as if it had been cut off with shears. He is made for a life of idleness; his hind legs bear no pollen baskets, so he could not fetch and carry if he would; his tongue is so short that he must needs eat from honey stored in a cell or be fed by his sisters, since he could not possibly extract nectar from a deep flower; nor is there any occasion that he should hang suspended weary hours for the secretion of wax, since he has no wax glands. His special accomplishment is his buzz, which is of extraordinary calibre and sonorousness. So fierce and loud is this, the song he sings when on the wing, that the novice feels inclined to retreat before him. But this music is undoubtedly meant to attract a queen to his vicinity, and is by no means a sound of menace; he is a burly, good-natured fellow, who is obliged to express himself in this rather coarse song. The term "good natured" is applied to him, not because we are certain that his temper is sweet, but because he has no means of expressing ill temper should he experience it, since niggard nature has given him no sting. He is always a clumsy chap, as awkward as his queen is graceful; but he can scramble out of the way with astonishing celerity when a murderously inclined sister attacks him. The time when he shows his princely qualities is when he is flying, for his wings are large and strong and carry him easily several miles if he needs to travel so far to win his lady. In his physical makeup he is a fine example of a purely feminine product; for the drone is a very perfect creature, even if he is reared from an unimpregnated egg. His magnificent compound eyes almost completely encircle his head, nearly meeting at the top, and thus crowding his simple eyes down into his "forehead." And such eyes as these mean something surely, for they are developed that he may be better able to see his heart's desire from afar. He also has marvellous antennae, the nine distal joints of which are completely pitted with smelling organs. The reason for this is that his queen has a fragrance all her own, sweeter to him than the attar of roses, and thus he is equipped, as Cheshire has proven, wath thirty-seven thousand eight hundred nostrils, in order to detect the perfume of her royal person at a distance. (Plate VII, *G.*)

The life-history of the drone after he hatches from his unfertilised egg is much like that of other bees, except that for him is provided a cell larger than that of the worker; he hatches from the egg about three days after it is laid, and during the week following he is carefully attended by the nurses who feed him on the rich chyle food at first; after four days they give him some undigested pollen, a food not granted to the larvae of the workers. At the end of a week his cell is sealed over with a cap that looks more like the crown of a derby hat than a cap, so spherical is it. Cheshire has shown that this cap is an especially fine example of engineering, being girdered by six struts of wax, the apex of the dome being not a skylight exactly, but rather a ventilator for the admission of air. (Plate VIII.)

It has been a question of much dispute whether the workers inspire the queen to drone-raising through building drone comb, or whether she takes the initiative in the matter. Certain it is the bees seem to love to build drone comb, perhaps because it is more easily constructed and requires less wax. It is also a fact that the queen prefers the worker cells, and in the spring or fall when there is little honey coming in, the queen will voluntarily pass drone comb, leaving it empty, and lay eggs in the worker cells, so she evidently knows her own mind. Sometimes when reduced to dire extremity the queen will lay worker eggs in the drone cells, but she does not do this unless the openings of the cells have been previously constricted by the bees. Sometimes also when the conditions are abnormal the queen will lay drone eggs in the worker cells and from these will be developed runty drones, which seem of little account. However, such conditions as these are very unusual.

When the drone is twenty-four days from the egg he cuts a circular lid out of the cap of his cell, and crawls out into a hazardous world. After a fortnight or so of moving about the combs and eating his fill he goes out of the hive and tries his wings. This he does on some pleasant day, about noon or a little after. As soon as he is sure of himself, he makes his flight longer, and the length of his journeys may only be guessed at. When he meets the queen they unite at once in the air, and after this they fall to the ground and she frees herself by tearing off and holding within herself the generative appendages of her dying consort.

In every hive are developed thousands of these princes royal, who are maintained at the expense of the colony until the dawn of that fatal day when the honey crop runs short; and then an unhappy experience lies before these useless brothers of the reigning house. Then their sisters chase them out of the hive apparently attempting to sting them, and, changed to furies, bite off their wings and harry them until they give up, great helpless creatures that they are, and fulfil their final destiny, which is to die for the sake of the colony.

Even the drone eggs, larvae, and pupae are not exempt on this appointed day of execution, but are ruthlessly killed, and their remains thrust forth from the hive. If conditions should change and more honey be made, a reprieve to the unhappy drone may be granted, for the length of his life is measured by the food supply. Any time during the summer when the bee-keeper finds the workers attacking the drones he may be very sure it means that the honey crop is exhausted.

Our pity is usually much more excited for the fate of the drone than for that of the busy worker, which dies from overwork at the end of a few weeks; and this is undoubtedly because the death of the worker seems voluntary, while the drones are manifestly murdered. Once we witnessed the slaughter of the drones in an observation hive, the entrance of which was too contracted to allow the bodies of the drones to pass. For several days the bloody-minded workers spent their energies in tearing their wretched victims limb from limb, and carrying them out in sections. Below a small crevice at the bottom of the hive we found a row of disjointed legs, wings and antennae from the mutilated drones, while the heads and broken bodies were thrown out of the front of the hive.

THE WORKERS. (Plate V, VI. VII.)

It is interesting to note that in the socialistic bee-community the work is carried on by unsexed females. It evidently has not been a part of the true economy of the perfect socialism to unite motherhood and business life in one individual; therefore, a division of labour takes place. The queen mother is developed into a highly efficient egg-laying machine, while all her worker sisters remain undeveloped sexually, and thus have time and energy to devote themselves to bringing up the young, keeping the house, getting the food, and administering the affairs of the body politic. Little wonder is it that the brain of the worker bee is much larger than that of the queen or drone, for she needs must exercise her mental powers far more than either. She is obliged also to pass through certain industrial stages in her development as a worker before she attains the full height of citizenship.

The life-history of a worker is usually as follows: The cell in which she is developed is the smallest of the comb, such as is ordinarily used for storing honey. She is not merely a fatherless creation, like the big drone, but hatches from an impregnated egg during the fourth day after it is laid by the queen mother. She is supplied with royal jelly, presumably the same as that which nourishes the queen larva, for about three days; afterward she is fed honey and digested pollen. This food is placed in the bottom of the cell, and the young larva floats in it and absorbs it through the body walls as well as through the mouth, which a little later she opens up pleadingly that it may be filled by the nurse bees. She grows like other infantile insects by shedding her skeleton skin as fast as she outgrows it; she does this with thoroughness, for she sheds the lining of the mouth, the gullet, the larger intestines and the tracheal tubes as well as the outside, this being a very thorough change of clothes, indeed; she does this about six times. Soon after she hatches she querls up in the cell, floating in her food, and at the end of four days' feeding she is a very fat, contented youngster. Six days from the hatching the nurse bees place over the cell, a cap which is made of wax and pollen, and admits the air freely. Then the young bee in the solitude of her own cell eats all the food that has been provided, spins about herself a cocoon of finest silk, which she weaves from a gland which opens in her lip; this is a very, very delicate cocoon, which remains in the cell as a lining, but so delicate is it that not until years have elapsed do the brood cells become contracted by these many silk wrappings of bees which have been developed in them. When the worker sheds for the last time her skeleton, she sheds the lining of the stomach and alimentary canal and all its contents, and changes to a pupa, which is a state of utter quiescence and during which wonderful changes take place in her anatomy. These changes which occur in the pupa are almost like new creation, for the legs, wings, antennae, and all of the other organs of the adult bee are developed from what was within the body of the footless, white grub.

Twenty-one days from the date of the laying of the egg, twelve days after the cell is capped, the worker bee sheds her pupa skin, pushes it behind her to the bottom of the cell, cuts a lid in the cap of her cell and pushes her way out, very likely after some friendly nurse has given her a little food to "stay her stomach." As she crawls out, she is a silvery-gray bee, as damp as if she had been out in a fog; no one hastens to greet her, or pays her the slightest attention, which is quite different from the case with the young ant, which is always fussed over and patted and petted by the nurses for some time after it emerges from the pupa skin. But the worker bee has to pat herself, and so she gives her face a rubbing, stretches and tries to straighten out her draggled clothing, and walks around trying to get acquainted as best she can with her sisters, who are too intent upon work to notice her. The first twenty-four hours of her life as a bee are spent orienting herself; but on the second day she learns to put her head down into the cells of unsealed honey and drink her fill. This is not a selfish and thoughtless act, for almost immediately she enters on her first duty, that of bee-nurse; and she must eat pollen and honey and digest them in order to make chyle for the bee brood, which she soon learns to care for most solicitously. It may be her lot to supply royal jelly to a queen cell and thus become a lady-in-waiting. In any event she very soon learns to be useful in many ways; she helps to build comb, and works very hard at capping the cells of the young bees when they are ready to pupate. She also helps to clean house if necessary, carefully removing all of the dirt and refuse at the bottom of the hive and dumping it out of the front door. During the extreme heat of the summer she must exert herself tremendously by fanning with her wings so that a draft maybe set up in the hive for the sake of the bees as well as to ripen the honey in the uncapped cells. During very hot weather, when the bees hang out, some of these young workers may be seen fanning "for dear life" on the outside of the hive. Having more zeal than wit, they dance a little glide back and forth, and fan as if they thought they were really accomplishing something.

The young worker usually takes her first flight when she has had her wings for about a week; she runs out on the threshold of the hive on some pleasant afternoon, and may be easily recognised, as she is callowness incarnate. She runs around a little, giving the impression of holding on with all her six feet as if scared, and then she lifts herself gingerly to see whether she truly can use her wings; then she circles around in great joy and learns to know well the place where her hive stands. About a week later she goes out into the wide world to seek her fortune and is likely to come back with a little load of pollen on her legs. When she comes back thus laden she buzzes around before alighting, and then rushes into the hive excited and delighted with her achievement, and as Mr. Root says so graphically, "tries her best to show off." Soon after, she becomes a staid worker and plays her part in the economy of the hive by bringing in honey, pollen, and propolis, secreting wax if need be, ready to defend her colony at the cost of her life, and so courageous that she as readily attacks a man as a mouse. Later it may fall to her lot to become executioner of her brother drones, or to devote herself to the queen and help lead out the swarm; or, by some mysterious election of the hive, she may be sent as a scout to find a proper home for her queen and her colony after they have swarmed. She is at the height of her powers and usefulness when about a month old, and at that time she will do any of the duties of the hive which she deems necessary, even to helping the young bees in the housework. She still has all her fur and her wings are as yet whole; but if there is much to do she is untiring and unremitting in her labours and, with never a thought of self, wears herself out. Her old age is evident by the loss of the fuzz, which was the pride of her youth, and the segments of her body become bald and shiny. Then her hard-worked wings begin to fray at the edges until there comes a day when, out on her quest for food for the colony, the broken wings and tired muscles refuse to support her, and she falls into the grass and dies; even then her last thought is not for self, but for the precious load which she struggles to carry home. Better thus for her to die in the field than to faint in the hive, for then do her vigorous sisters seize her and thrust her forth, and she falls into the refuse heap in front of the home, which she has so eagerly wasted her life to sustain. There is no gratitude and no pensioning in the bee-world;

death and oblivion are meted mercilessly to the most ardent workers when they fail, for thus, and thus only, can the colony be kept strong. The individual is nothing in the perfect socialism, and the colony is everything; the treatment is Spartan, with none of the weakness which makes us keep alive the hopelessly insane, the idiotic, and the criminal.

THE LAYING WORKER

When the colony is queenless, worker bees may develop the ability to lay eggs. As they have never mated, they lay unfertilised eggs, which develop into drones, and thus stock the hive with these royal cumberers of the commune. It is interesting to note the difference in prejudices that obtain in the hive and in human society. In the latter we regard it as scandalous when the female, avoiding the duties of motherhood, goes abroad gathering honey and pollen at her own sweet will; but in bee society it is not merely a scandal, but a misfortune, when the worker bee has ambitions to be a mother. The laying worker is a bee gone wrong and a menace to the colony. At the same time she is a nuisance to the bee-keeper and great may be his tribulation before he is rid of her. As might be expected, she does not do her work well; she usually does not lay her eggs in regular order, as does the queen, but scatters them here and there and everywhere, and is quite likely to fasten them to the sides of the cells, instead of to the bottom. She lays her eggs sometimes several together in worker as well as in drone cells.

As a laying worker looks like any other worker, it is useless to try to find her. However, her presence may be detected by the irregular appearance of the brood, and especially by the high drone caps on the worker cells and, finally, by a superabundance of drones. To meet this difficulty, an ounce of prevention is worth several pounds of cure, and great care should be taken to prevent colonies from becoming queenless. In case, through carelessness, a colony is thus victimised it will usually refuse to accept a queen, though sometimes it may be induced to accept a capped queen cell. If this is not successful, the combs, with the bees adhering, should be removed to an empty hive nearby, placing a frame of brood containing a queen cell, if possible, and a frame or two of foundation in the old hive. The workers, coming back from the field, will enter their hive and the moved comb will soon be deserted by all except the laying worker; she, with her characteristic fatuity, will remain on the deserted combs, laying eggs until she dies of exhaustion. A surer remedy than this, but a more troublesome one, is to unite this colony with another, or to scatter the combs from the victimised hive, bees and all, among other colonies of the apiary; meanwhile giving the depleted hive a frame or two of good brood, with a queen cell, if possible, so that the bees that return to it will find normal conditions. What happens to the laying worker when she finds herself in a colony with a queen, we do not know. Probably, if she persists in laying eggs, she is killed; possibly she forsakes her evil ways, and returns to the straight and narrow path of respectable citizenship.

We do not understand why laying workers are developed. Some have claimed that too much royal jelly was given them when larvae; and some, that after a colony is queenless, the jelly is fed to workers and thus develops them so they are able to lay eggs. They appear among the Cyprian and Syrian bees more frequently than among the Italians.

RACES OF DOMESTIC BEES

Several races of the honey bee have been developed in different countries. Some of these have been imported into the United States, and many experiments have been made to determine their relative values.

There were no native honey bees in North America north of Mexico, and the black or German bees were the first to be brought to this country by the pioneers. The wild bees which stock the woods of our country to-day are chiefly black bees, the descendants of swarms which have escaped from apiaries. For many years the black or German race was the only kind in general use here. Within recent years the eastern races of bees, Cyprians, Holy-Lands, or Syrians, and also the Egyptians and Carniolans and Italians have been introduced. Of all these, only the Italians have come into universal favour.

The Italians are the classic bees which were discussed by Aristotle, and sung about by Virgil, who describes their bodies as "shining like drops of sparkling gold." The Italian worker has five yellow bands that mark the front portions of the five segments of the abdomen which lie next to the thorax; the two posterior bands are made by yellow hairs and are therefore likely to disappear as the bee gets old and bald. But the three front ones are made not only by yellow hairs, but also by the yellowish transparency of the front part of each of the three anterior segments, or body-rings. As the segments of the abdomen telescope, more or less, these three yellow rings may not always be visible. Mr. Root's test is to feed the bee with honey until the abdomen is distended and place her on a window pane. If three distinct translucent bands can be seen, the insect is a pure Italian. If only two bands are evident, she is a hybrid.

In comparison with the Italians the black bees are inferior in many particulars. Their only superiority is that their honey in the comb is a trifle whiter, and they are more easily shaken from the frames than are the Italians, and thus are sometimes preferred by the man who works for extracted honey. Their points of inferiority are their nervousness and irritability, their tendency to rob, their inability to cope with the bee-moth, and the difficulty with which the queen may be found.

The Italians are far more "civilised" than are the black bees, and seem willing to credit the operator with good intentions; they can defend themselves better from pests, their queens are more prolific and, on the whole, they are more industrious than the blacks, and having longer tongues than the blacks, they can get nectar from a wider range of flowers.

The hybrids are the result of crossing the Italians and the blacks. They are likely to be excellent honey-gatherers, but unfortunately they usually inherit the irritability of the blacks, which they express with the strength and energy inherited from the Italians. Therefore, they are not looked upon with favour. These unwelcome hybrids are likely to appear in any apiary, for, however pure the Italian queen, she is likely to mate with a black drone in almost any locality.

CHAPTER V

THE INDUSTRIES OF THE HIVE

THE SECRETION OF WAX

VERY little do we know of the mysterious process of wax-making. The interior of the bee is a chemical laboratory where no visitors are allowed; at best we have been obliged to stand outside and guess at the formulas used within. We know that honey enters largely into the composition of wax, and that the bees when secreting wax usually have pollen in their stomachs, although Huber and Cook have both demonstrated that bees make successful comb when deprived of pollen, and when fed on sugar syrup instead of honey. But to make this experiment of much value the bees must needs have been deprived of pollen all of their lives instead of a few days. It seems to be generally conceded that nitrogenous food is needed for the best product of wax-manufacture, though nitrogen does not enter into the composition of the wax itself.

It is variously estimated that it requires from ten to twenty pounds of honey to produce one of wax; so it is apparent to even the novice in bee-keeping that wax is a very expensive product. One of the

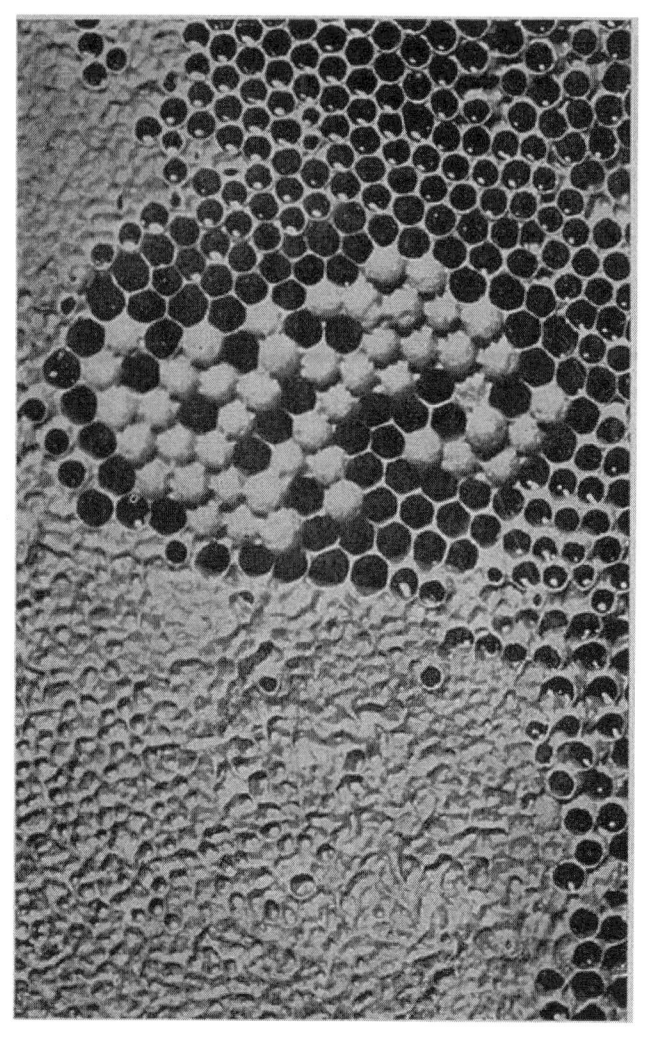

Photograph by Brown Brothers

PLATE VIII. DRONE CELLS IN A COMB OF HONEY

Honey capped over; drone cells bulged out; upper cells partly filled with honey but not capped over.

Photograph by W. Z. Hutchinson

PLATE IX. Side of hive removed, showing bees at work; comb being made without artificial foundation.

results of the process of wax-making is the elimination of oxygen from the honey. There is of weight eight times as much oxygen in honey as of hydrogen and carbon combined; while in wax there is at least sixteen times as much carbon and hydrogen as of oxygen. Though wax is a fatty substance, yet it is not the animal fat of bees, as is so often asserted; it is a product especially developed for a far different purpose than is the fat of animals. The bees are much superior to us in this respect, since they manufacture from their own bodies the building materials for their homes.

The special apparatus for the secretion of wax is very interesting to the student skilled in microscopic investigation. The outward or visible portion of this apparatus consists of four pairs of little membranous plates on the under side of the abdomen; these are not visible unless the body is torn apart and dissected, because they are on the front portions of the second, third, fourth and fifth abdominal segments, and each is covered by the rear portion of the segment just in front of it. Immediately within each of these wax plates is a gland which secretes the wax in liquid form, and it passes through the membrane by a kind of osmosis, considerably more mysterious than is that most mysterious process ordinarily. As the wax passes through the membrane it hardens and is pushed backward behind the segment which covers it and protects the wax plate, and appears as a pearly scale on the abdomen of the bee. (Plate VI.)

The wax glands, when studied by the histologists, are found to consist each of a specialised area of the layer of cells that form the active living part of the body-wall of the insect. When active these cells are much thicker than the corresponding cells in other parts of the body-wall; but if examined during the winter, they do not differ greatly in appearance from other cells of the hypodermis. (Plate XXV, Fig. 5.)

When wax is needed, a certain number of self-elected citizens gorge with honey and hang up in chains or curtains, each bee clinging by her front feet to the hind feet of the one above her, like Japanese acrobats; and there they remain, sometimes for two days, until the wax scales appear pushed out from every pocket. It is not hard to understand that, since much honey is needed for the manufacture of wax, a bee after filling with the raw material would produce much more wax by keeping quiet than by using any of the gorged honey for energy in moving about and working. But the necessity of "holding hands" while this work goes on must ever remain to us another occult evidence of the close relations of the citizens in the bee commune. (Plate X.)

While most of the wax is produced from these quiescent suspended individuals, yet any bee-keeper who is observant has discovered that at the height of the honey season many of the workers coming in laden from the fields will have wax scales protruding from some or all of the pockets. We once captured one of our bees, working on a white clover blossom, which had six of these wax scales ornamenting her abdomen, and which proved her a bee of resource, since she was able to work and make wax at the same time. However, there is a choice about the wax-making. It is no willy-nilly production caused by gorging with honey, for it is never made except when the colony needs more comb.

THE COMB-MAKING (Plate IX.)

It is often stated that after the wax is secreted and pushed through the wax pockets the scales are removed by the wax shears on the hind legs of the producers, and are passed forward to the front claws, and then thrust into the mouth; here the wax is warmed and perhaps chewed with saliva and made malleable, thus somewhat changing the chemical composition and fitting it to be moulded into comb. But there is a hiatus in our knowledge just at this point as to whether the bees which secrete the wax take it off and make comb, or whether other workers harvest wax-scales from the suspended individuals and mould them into shape; or whether perhaps the scales fall from the suspended "curtain" to the bottom of the hive and there are gathered by the ever-busy young workers. Professor Kellogg, who studied bees in an observation hive, is inclined to think that all of these methods are used, while Mr. A. I. Root describes the process graphically thus:

> If a bee is obliged to carry one of these wax scales but a short distance she takes it in her mandibles, and looks as business like with it thus as a carpenter with a board on his shoulder. If she has to carry it from the bottom of the honey-box she takes it in a way I cannot explain better than to say she slips it under her chin; when thus equipped you would never know she was encumbered with anything, unless it chanced to slip out, when she will dexterously tuck it back with one of her forefeet.

Honey-comb has been the delight of mathematicians from the earliest ages. The plan on which it is built, if perfectly carried out, would be the incarnate perfection of strength and space for holding fluid contents. This fact so delighted the earlier mathematicians that they set to measuring the angles of the cells and their pyramidal bases, with truly wonderful results. But with the later methods of exact measurement it has been demonstrated that the cells are rarely perfect in construction; and that the angles, as well as the faces of the rhombs on which they are built, vary. Because of this there have been developed doubters and pessimists who declare that honey-comb is the result of chance; and that cells, crowded together, must, from the nature of things, become six-sided; and that bees are not mathematically wise. With this conclusion we do not agree in the least, although we admit that the fortuitously six-sided cell may have been a step in the education of the bee-artisans. But we would ask the pessimists to explain why, if all is chance, the bees build so perfectly the central part of the comb which forms the bases of the cells. This central part is built first and is fashioned of rhombs, which are made into alternating three-sided pyramids. Who dare assert that reasonably perfect, alternating rhombic pyramids are fortuitous! The fact that the combs are rarely perfect in construction proves naught against the mathematical prowess of the bees; it simply proves that the bees are a practical race, and not bigoted, and are therefore unwilling to sacrifice everything for the sake of precision. The construction of their waxen cells is for economic purposes rather than for proving mathematical formulae. Honey-comb shows how economy of room, building materials and mathematical theories may coincide, and shows also that the bees have taken advantage of the fact. Some of the savants have asserted that the rectangle or the equilateral triangle would have been quite as efficient as working plans for constructing cells for storing honey. But probably the bees, originally, made their cells to fit their brood and would not thus build a cell which would surround a larva with unfilled corners. The hexagonal cell was better fitted for their needs, so they developed it.

After a piece of the central portion of comb has been constructed the bees begin, usually, at the centre and pull out the sides of the cells from the foundation. Experiments in coloured foundation shows that this may, if thick, be pulled almost to the margin of the cell. This is why bees so readily utilise machine-made foundation; they pull out the edges of these pressed combs and thus save themselves much labour in wax-making. The worker-cells are a little more than one-fifth of an inch in diameter, and a little less than one-half an inch deep; the drone-cells are a little more than one-fourth an inch in diameter and a little more than a half-inch in depth. It is interesting to see the comb which has in it both worker-and drone-cells, and note how the transition is made; the two sizes are harmonised by a row or two of cells that are irregular. Honey is stored in both drone-and worker-cells, usually in the latter; although our bees seem to have a fondness for making drone-cells for storage. When the bees begin to cap a cell, they commence at the outside and work toward the centre. There is not a prettier piece of engineering anywhere than the cap of a honey-cell, with six little girders extending from the angles of the cell and holding the flat cap at the centre. Honey is capped with wax, but brood is capped with a mixture of wax and pollen, which admits air. Though the cell-walls may be thinner than .0018 of an inch, comb is wonderfully strong, and may weigh one-twentieth or less than the weight of the honey stored within it.

An interesting fact about the manufacture of comb is that no one bee constructs a cell and no one bosses the job. A bee will come along with a little wax and put it in place at the side of a cell, and then will run off and do something else; another bee passing sees this bit of unfinished work, gives it a few pinches and polishes it a little, and then does something else. Several bees may thus lend a mandible before the cell is perfected. Any bit of comb-building seems to be the result of a consensus of public opinion and not of individual skill and enterprise. There is a oneness in bee enterprises which harmonises capital and labour, and which precludes strikes and lockouts.

THE PRODUCTION OF HONEY

In trying to fathom the mysteries of honey-production, scientists have dissected the bee with greatest care; but they have usually been obliged to guess at the uses of such organs as they could not understand by analogy. To-day, after all the excellent work of investigators, the process and formulæ of honey-making remain unrevealed.

The nectar, when taken from the flowers by the bees, is received in a special reservoir, called the honey-stomach. It is supposed that the secretion from the glands in the head and thorax is added to the nectar as it is swallowed, and induces the chemical action which, in the honey-stomach, changes the cane-sugar to the more digestible grape-sugar, and brings about the other changes that finally result in the production of honey. The chemical composition of honey varies, perhaps for two reasons: It may be more perfectly digested sometimes than at others; and the nectar of different flowers may vary chemically. However, all honey contains water, glucose, a small amount each of albumenoids, mineral matter, essential oils and formic acid. While most of the chemical changes take place in the honey-stomach of the bee, yet the honey is made perfect by ripening in the cells; these are left uncapped for a period of time and the current of air, always in action in the hive, evaporates the water and thus thickens the honey.

Ignorant people believe that honey is regurgitated from the true stomach of the bee, which is far from true. The honey-stomach is simply a reservoir in which the honey is made and then stored until the bee can empty it into a cell. Her true stomach lies behind the honey-stomach and connects with it by a mouth that can be opened or closed at will; when she wishes to eat some honey, she opens the stomach mouth and takes in what she needs. The chyle which she manufactures in her true stomach to feed the young bees, when regurgitated, does not pass through the honey-stomach; instead, the mouth of the real stomach is pushed up through the centre of the honey-reservoir until it connects directly with the æsophagus. (Plate XXVI, Fig. 7 *p.*)

When an ancient Roman was asked on his hundredth birthday how he had preserved his vigour, physically and mentally, he answered laconically, *Inerius melle, exterius olea*—"Inside with honey, outside with oil." He spoke wisely, for honey is undoubtedly the most healthful of sweets, because it is so largely composed of the predigested grape-sugar.

It is hard for us to realise that, until comparatively recently, honey was the only sweet in general use. Cane-sugar was not commonly eaten in Europe until the seventeenth century, and previous to that time honey held sway as the sweetening medium of all foods. The amount of honey produced in the United States now is estimated to be more than 125,000,000 pounds per year, which shows that it has retained its value as a food, though it must compete with cheaper cane-and beet-sugars. It still remains the most wholesome and digestible of all the forms of sugar, and should be used even more generally than it is at present.

THE MAKING OF BEE-BREAD

Flower wisdom is scarcely appreciated by those who deem all wisdom the product of consciousness; but if wisdom may be attained through the experiences of living and overcoming difficulties, then there must be such a thing as flower wisdom. Otherwise there would not have been such a prodigal production of pollen that a tithe could be spared for the bees, to induce them to become common carriers of the flower world. Many blossoms which do not secrete nectar pay their taxes in pollen, the bread-stuff of the bees, while others pay in both commodities.

A bee when gathering pollen for food collects it with her tongue and forelegs, mixing it, perhaps, with nectar or saliva so it will hold together. It is cleaned off the tongue and front legs by the middle and hind legs, and by them packed in the pollen baskets on the tarsi of the hindlegs, and moulded there into great golden balls. Little wonder that the ancient Greeks, noticing bees thus laden, and consequently flying low, declared that the bees of Hymettus tied pebbles on their legs to weigh them down. (Plate VII, *A, B.*)

When the bee arrives at the hive she selects, usually, a worker-cell and, backing up to it, thrusts in her legs and scrapes off the pollen by a dexterous movement like that made by a cook scraping dough off her hands. The bee bringing the pollen considers her duty done in furnishing the flour, and leaves the bread-making to one of her younger sisters, who is devoting the day to domestic duties. Needless to say, bee-bread is unleavened; it is made by the very simple process of packing the pollen firmly into the cell, the utensil employed being the head of the bread-maker, which she uses cheerfully as a mallet for this purpose.

Bee-bread is necessary as a food for young bees and admirably supplements honey in its composition, being rich in albumenoids and nitrogen. To our taste it is rather bitter and disagreeable, as those of us can attest who ate comb-honey from the hives of old, before movable frames and supers were generally used. However, under the new régime, it is rarely placed in the sections of the supers, but sensibly stored in the brood-combs, near where it is used, and thus seldom appears upon the table.

THE PROPOLIS, OR BEE-GLUE

Though bees are most successful manufacturing chemists, yet they are not above using ready-made substances if they find such to their liking. Thus, propolis is not produced by bees, but is gathered by them from various sources, and is used as a cement and a varnish. Certain trees and smaller plants protect their buds in winter by resinous coats; and it was quite like the adaptable bee to find use for this resin in her own domicile.

The elder Huber, whose observations of a century ago have been verified, discovered the source of propolis; he planted poplars in pots and placed them near the hives, and the bees were seen in the act of collecting the resin from the buds. They have been observed by others, working on the buds of the horse-chestnut, birch, willow, alder, and even the balsam fir. However, the bees have no prejudices in favour of any kind of resin, anything will do so long as it answers the purpose; hence they visit shops where furniture is being finished and appropriate the varnish without saying "please." And Darwin mentions the fact that bees collected a cement of wax and turpentine, used to cover trees from which the bark has been removed. If any old hives or fixtures with propolis on them be left around the apiary, the bees make all haste to save every particle of the precious stuff.

One of the oldest superstitions about bees is that they will gather on the coffin of their dead master; and authenticated instances of the kind are on record. But this beautiful tradition is made empty of sentiment by the assertion that the bees assemble there not to mourn their dead master, but to gather the varnish from the coffin. Some iconoclasts ascribe to this the origin of "telling the bees" when some member of the family dies; but we believe this beautiful custom originated before varnished coffins were in use, and was the natural outgrowth of the close relations of bees and the home for many centuries.

The bee collects the propolis by cutting it off with her mandibles and packing it in her pollen baskets; and when she arrives at the hive, she never attempts to unload it herself, evidently deeming it safer to let the sister, whose business it is at the moment to stop cracks and crevices, take it from her legs and apply it at once.

There are various uses in the hive for the bee-glue: It is used as a filler to make smooth the rough places of the hive; it holds the combs in place; it calks every crack; it may serve as a sarcophagus for any intruder too large to be pitched out: snails and slugs have been found thus encased; it is applied as a varnish to the cells of the honey-comb if they remain unused for a time; and it is especially useful as a window shade. A nature-study teacher of our acquaintance established an observation hive in her school-room, which had an uncovered pane of glass set in one side so that the pupils might observe the interesting life of the hive. To her dismay the bees straightway hung a curtain of propolis over the window and so shut out intrusive eyes.

One of the chief uses of propolis is to try the temper of the bee-keeper. If there is the least crack between hive and super, or cover, the two are glued together so that nothing but a knife and much muscular force can loosen them. Cover blankets are stuck fast; the frames are welded to their places by it,

Photograph by W. Z. Hutchinson

PLATE X. Bees hanging in a curtain secreting wax; comb being built in the centre above.

Photograph by Verne Morton

PLATE XI. HIVING BEES
Cutting down the swarm.

and it plasters into immobility all the modern appliances of the hive which man invented to be movable; and section-boxes are so stained with it that they have to be scraped before being sent to market. Above all, it gets on the hands of the operator, so that the digits of each act as one, or not at all, and everything touched sticks to them as if they were magnets; it also daubs the clothing, and if profane men were ever bee-keepers, it would incite them to wicked remarks. However, alcohol, applied to hands and clothing, solves the difficulty by dissolving the propolis; and a bottle of it is a most necessary adjunct to the equipment of the apiary. Benzine, gasoline, ether, and chloroform are as efficacious as alcohol, but not so pleasant to use. Boiling lye will clean clothes and apparatus that have become clogged with bee-glue.

CHAPTER VI

THE SWARMING OF BEES

WHY BEES SWARM

In case of most of the higher animals, including man, the species are spread over the face of the earth through the simple plan of "marrying off" the younger individuals, and allowing them "to set up for themselves." As there is no individualism, no indulging in love or hatred for each other among the perfect socialists, this plan is manifestly not at all adapted to provide for the increase of social insects where the colony acts as an individual. The bee plan is as follows: When a colony gets large enough and strong enough it divides into two. As the Amœba, the simplest of animal organisms, divides the individual in order to multiply its numbers, so the bee-colony as an organism divides and multiplies. It is interesting that the method of increase characteristic of the lowest order of animals should repeat itself among beings so highly organised physically, when the individuals merge into a social unit. This analogy might well give the social philosophers an opportunity for some real thinking. Much discussion has found place in human annals as to whether the queen or the workers take the initiative in this hegira of the bees, and much evidence is advanced on both sides of the question. But it is a profitless discussion for us to indulge in; the more we study bees, the more firmly we are convinced that we know little of the forces which govern the bee body politic. We only know that at certain seasons of the year when the successful colony has plenty of brood and honey, the old queen and many of the old and experienced citizens, go away from the hive and form a new colony elsewhere, leaving the young queen and the younger bees in possession of the homestead, thus reversing the human custom. While in the air, or when clustered, the swarm looks sufficiently large to be composed of the whole colony, yet the swarms are smaller than they look, for rarely more than 20,000 bees go away with the old queen.

WHEN BEES SWARM

The swarms usually come off in June in a climate like that of New York State. Some strong colonies may swarm in May if the season is early, thus making us glad; for the old verse, "A swarm of bees in May is worth a load of hay," is based upon practical experience.

The when and the immediate why of the swarming of bees are closely connected. There are several conditions which conduce to swarming: the presence of young queens ready to issue from the cells; the crowding of the hive with brood, bees, and honey; the presence of too many drones; the extreme heat of an overcrowded hive on a hot day; this latter is a most cogent reason, and it is well for the bee-keeper to shade his hives to prevent this. The swarms are likely to issue between 9 A. M. and 2 P. M., although enough swarms come off at unseemly hours to make any rule valueless except as a prophecy. (Plate XI.)

HOW TO HIVE A SWARM

"The bees are swarming!" These are magical words, which arouse every member of that family whose pride is a few hives in the garden. It is a cry that starts the sluggish blood, and sends a quiver of excitement up and down the spinal column while one rushes to the scene of action. How gracefully that moving mass of black particles undulates in the air, as if it were a drifting cloud instead of a self-willed, one-minded colony of socialists! How the heart rises and sinks inversely to this rise and fall, and how hopeless it seems when the swarm lifts itself superbly over all surrounding obstacles, and disappears above the tree tops! No one who has had this experience will wonder at the ancient custom which obtains even now in the country districts on such occasions of beating tin pans, ringing bells, and shouting "whoa" at the top of the lungs. All of this racket had its inception in the needs of the bee-keeper to adequately express his feelings at this crisis. If the bees ever stopped and settled because of this din, it was probably from sheer amazement at witnessing such folly on the part of human beings; this explanation would hold, perhaps, if bees ever evinced any interest in human beings, except when they obstruct the bee-path. However, most of our ancient customs were founded in utility, and it is not likely that this traditional pandemonium would have been practised for centuries without some reason. Mr. Root, who thinks before he speaks, suggests that the swarm follows the queen and the scouts through listening to their song, that of the queen being easily distinguished from the hum of the workers when on the wing; and that it is quite possible, therefore, that the noise, if loud enough to drown the voice of the queen, would cause confusion on the part of the flying bees and a consequent settling. But from what we remember of our own early experience, we are convinced that the bees were less confused by the noise than were the people engaged in making it.

The next most widely practised of the ancient methods is that of throwing dirt on the swarm to stop it; this certainly is efficacious if there happens to be enough loose earth at hand. However, throwing dirt is a reactionary performance in the physical as well as the moral sense; and usually the bee-keeper who throws dirt at a soaring swarm must needs stop soon to get the motes out of his own eyes before he is able to see where the swarm has alighted.

In these enlightened days everyone who has a cherry tree or an apple tree, a currant bush or a potato patch, is sure to have a fountain pump of some sort; and never is this instrument a greater boon than when it throws a fine spray upon an absconding swarm of bees. It brings them to a stop very soon; it may injure the feelings of the bees, but it certainly does not injure them physically, as it simply impairs their power of flight by wetting the wings. Even after a swarm has settled, a little sprinkle of water will keep it clustered safely until the hive is made ready to receive it.

It is highly desirable that the swarm should cluster on the tip of a branch not far from the ground; for then the process of hiving is comparatively simple. My personal plan has been to place a sheet, kept for the purpose, on the ground near the clustering swarm; place on this a covered hive filled with frames containing brood foundations. Lift the front edge of the hive about an inch by putting blocks under the two corners; then cut the branch above the cluster, and taking it in hand shake the bees off in front of the hive and placidly watch them hive themselves with true bee celerity. This use of the sheet is a habit formed in childhood, and I persist in it, though my partner derides the practice. He shakes the bees down on the board on which the hive is standing; or he takes the top off the hive, and shakes the bees down among the frames in the most summary fashion. But I think the sheet makes a softer mattress for the little citizens to fall upon, and certainly they find their way from it more easily into the hive. I am a conservative person, and like to do things as I always have done them before—a conservatism that is by no means dangerous in our apiary where the senior partner is given to new ways and many inventions.

If the bees alight high in a tree, then our methods have been to get at them by climbing the tree, or attaining the branch by the use of a ladder. However, I doubt if there was ever a fruit found on any tree that needs quite so much care in the picking as does this; and it is decidedly a ticklish performance to clamber down a tree holding gingerly a branch laden with a swarm of bees in one hand and clutching at supporting branches with the other. Sometimes the bees are not accommodating enough to alight on the end of a branch that may be cut off. They may even go so far as to cluster on the large branch itself; then there is nothing to do but to brush them off in a box with a bee brush, a performance which they object to; or to dip them off with a tin dipper, or to jar them into a basket, and then to dump them out in front of the hive. The most embarrassing situation of all is when the swarm clusters on a tree trunk. Squaring the circle is not a much more difficult feat mathematically than to brush all the bees into a square box from this cylindrical position. It is usually necessary to bring the smoker to help in elucidating this problem; for, paradoxical as it may seem, smoke properly applied clears up many a situation in the bee business.

If only one were able to find the queen in the clustering swarm and secure her by placing her in the hive, the work would be easy, for the other bees would soon follow. But to hunt for the queen in the clustering swarm is, for most of us, quite like hunting for the traditional needle in the haystack; and while we are hunting we are wasting precious time, as the swarm may make up its collective mind to remove from our vicinity its component atoms. In shaking, or dipping, or brushing the bees into the hive, we should remember to deal with them as gently as the situation and mental perturbation will permit.

Sometimes the bees seem to feel defrauded at having a house chosen for them, and insist upon swarming out of their new quarters within a day or two after they are placed in possession. Thus it is a wise precaution to give the hives of the newly swarmed attention for two or three days, lest they indulge in this sort of perversity. If the bees seem unsettled and unhappy, and hang around home instead of going into the fields to work, it is advisable to place in the middle of the hive a frame of unsealed brood taken from some other colony. A bee's motherly instinct may always be depended upon, since caring for the young is a shining civic virtue as well as a domestic duty among the bee people. When there are helpless larvæ to care for, the bees willingly forego every personal inclination to pack up and move, and cheerfully proceed to give the youngsters every attention.

DECOY HIVES

When the self-appointed scouts of a bee colony start out house-hunting for the swarm which is about to issue, they evidently examine the premises carefully; if they find a house for rent in the immediate neighbourhood, they are likely to go no farther in their quest. Thus it is the practice of many bee keepers to place about the apiary hives empty, except for brood-frames filled with foundation, hoping thus to entice the swarms to take possession, and save all trouble. Mr. Root goes so far as to advise that such hives be fastened in the lower forks of trees in the neighbourhood of the apiary, and thus provide a most expedient bee tree. A decoy hive should not contain more than three brood-frames, as other frames may be added when the bees move in. The reason for this is that the sheets of wax foundation, when sufficiently close together for the convenience of the bees, prove also entirely convenient for the occupancy of the bee moth; and one ought to be particular about one's tenants when renting houses in the apiary.

Mr. West's device for saving swarms is the most alluring of any about which we have read. Since the clipped queen cannot fly, she expresses the aspiration within her breast by climbing anything at hand, like a blade of grass or a shrub. Mr. West observing this, drives a bare, forked branch into the ground a few inches in front of the hive, having cleaned the ground between it and the hive of all obstacles. This branch having a few twigs upon it, is leaned away from the hive entrance. The queen promptly climbs this tree, like Zaccheus, and the swarm clusters around her and remains there conveniently at hand for hiving.

MAXIMS FOR HIVING BEES

Clip the old queen's wings.

Go through the hives every ten days to destroy queen cells.

As the swarming season approaches, have hives ready with foundation in brood-frames, and hive-stands ready to receive them.

Keep a serene spirit while hiving bees.

The hive in which the colony is placed should be kept cool and not heated from standing in the sun; it should be shaded, as the swarm will not enter a hot hive.

If the bees refuse to accept a hive, give them a frame of unsealed brood to reconcile them.

CHAPTER VII

HOW TO KEEP FROM KEEPING TOO MANY BEES

THE PROBLEM

"Aye, there's the rub!" For the beginner who wishes to keep a few bees, this is the most difficult problem to solve on the bee-keeper's slate, and it must finally be solved by each according to his own capacity and method.

We confess frankly that we were once driven out of the bee business because we were too successful. Caring for fifteen or twenty hives was a delightful avocation. We kept our colonies strong, for we wished to make comb-honey; consequently, splendid swarms came off, and we had the fatal gift of seeing them when they came and of hiving them successfully. Thus our avocation began to intrench upon our vocation in a most alarming manner. While we enjoyed taking care of our bees, we were nevertheless following the vocation which we had chosen, and which we liked best; and the time came when we were obliged to decide whether we would leave our regular business and become bee-keepers, or abjure bee-keeping and attend to our regular business.

We gave away what swarms we could and we sold some; but selling bees is a business quite as much as caring for them, so that was not practicable. Philosophically, we argued that since we had enough bees we would let the swarms that came off abscond and bid them God-speed; but here we reckoned without properly considering the amount of human nature which had fallen to our share. Although we knew that every swarm in our apiary above twenty would be an embarrassment and a tribulation, we could never rest content not to hive a swarm when it issued; and the more unattainable the place where it clustered and the less we wanted the bees, the more determined we were to secure that special swarm.

Such inconsistency brought its own punishment, and our only resource was to sell out; and for many years the spot in our garden which our apiary once occupied was never viewed without a sense of loneliness and longing for its busy little tenants. During these years we thought it over and finally came around to the right frame of mind, and firmly concluded to keep bees for our own honey and our own happiness. We limited our ambitions to ten hives; and last summer when our best-tested Italian queen took us unawares and departed with a large and enthusiastic following, we did not mourn; all we did was to venture to hope that the young queen left in the hive had mated with one of our own handsome drones, and not with a mad black prince from one of Mr. Coggshall's take-care-of-itself apiaries in our neighbourhood. As soon as our new brood made us sure that our queen had made no *mésalliance* we were entirely content.

Our lack of success in preventing swarming when trying to produce comb-honey was a source of great chagrin to us until we read that so eminent a bee-keeper as Mr. Hutchinson declared that "there is no way of preventing first swarms profitable to the comb-honey producer," and then our feelings were salved. The following are in brief a few of the more successful ways practised to prevent increase:

By clipping the queen's wings.—Almost all bee-keepers practise this now, whatever their method of preventing increase or securing it. A queen with clipped wings is necessarily a "stay-at-home body," and the swarm will not leave without her. However, when depending upon this method it is very important to guard against the hatching of new queens, and this can only be done by closely scrutinising the brood-comb to discover and destroy the queen cells. The brood-frames should be examined in each hive about once a week during the months of June and July, if this method is to succeed. Many a time have we sat smilingly by and watched a swarm come out of the hive with great pomp and circumstance, only to sneak back when it was discovered that her majesty was unfit for travel.

By the use of a queen-trap.—This is a device used by some instead of clipping the wings of the queen. It is a box of perforated zinc placed over the entrance of the hive, the slots in it large enough to allow the workers to pass in and out, and small enough to hold back the queen and drones. Although this method saves time, yet comparatively few recommend it. The cost of the trap is one objection; but the greater objection seems to be that it inconveniences the workers when returning from the fields; and is, therefore, likely to affect the amount of honey stored, since much time is lost and some annoyance occasioned to the bees by being obliged to squeeze through; it also scrapes the pollen off their legs. (Plate XVI.)

By giving-room.—Supposing that our queen is clipped or confined, which is the first step, the next is to give plenty of room in the brood-chamber. Lack of room for brood and honey is one of the most potent reasons for inducing swarming. So the first thing to do after the disappointed swarm comes back to the hive is to tier up the supers. It is also well to remove from the hive a frame or two of brood which may be put in the hive of some weaker colony; in place of the removed frame is substituted another containing a starter of foundation; and the would-be swarm, finding that there is plenty to do, is content to remain at home for a time.

By extracting honey.—It will often pacify a colony to take the frames from the brood-chamber and extract the honey from them. This may be done when the brood is present, if care is taken not to run the separator so rapidly as to throw out the larvæ, a performance quite as distasteful to the young bees as to the consumer. However, this should not be practised unless much honey is coming in, as otherwise the brood may be starved.

By using large hives.—Many bee-keepers of high standing have practically solved the swarming problem by using large hives. The Dadants, well known on two continents as successful apiarists, use the large Quinby hives in their own apiaries, and have introduced them into France and Switzerland. The Dadant-Quinby hive has about the capacity of a twelve-frame Langstroth. The frames are both deep and large, measuring 18½ by 11½ inches, and so give the queen plenty of room. The Dadants have no trouble with swarming, as only enough swarms come off to make good the winter losses in their apiaries.

There are three reasons why we have not used these large hives: first, they are too heavy to handle well, being altogether too productive of backaches. Second, they are necessarily much more expensive, as wider pieces of perfect lumber are used in the making. Third, and most important to us of all, we find it difficult to produce comb-honey in a large hive; when the bees have so much room in the brood-chamber, they discover no reason for carrying honey up into the supers. If we made extracted honey, as do the Dadants and the European apiarists, we would certainly use the larger hives, simply to be rid of this nuisance of constant swarming. The colonies grow to be so much stronger in the larger hives that they are much better able to withstand the vicissitudes of winter than are smaller colonies, which is a great advantage.

By the brushing or shaking-out method.—When the bees at the beginning of the honey-flow seem to be getting ready to swarm, the hive is moved to one side of the stand and on the exact site is placed another just like it, which contains frames set with foundation starters. As gently as possible the bees are shaken or brushed from the frames of the old hive upon the threshold of the new, great care being taken to include the queen. The supers from the old hive are then placed upon the new hive with a queen-excluder between. The old hive may stand beside the new one until the brood has emerged, when all of the inhabitants of the old tenement are shaken in front of the new habitation. This shaking of a colony into a new hive so surprises and confounds the bees that they get the impression that they have already swarmed, and go to work with all diligence in their new quarters. The partly filled supers from the old hive encourage them mightily in the ways of well-doing.

By dividing swarms.—This method we have practised with quite satisfactory results. The troublous question is just when to divide. If we divide too soon we weaken the colonies, and decrease the honey harvest. If we wait until too late, the bees do the dividing themselves. The process is as follows: The queen's wings are clipped before the new queen is to emerge, and she is placed in a new hive furnished with brood-frames containing foundation-starters; enough of the bees are taken to the hive with her to start a good colony, and the deed is done. However, the fact remains that in this way the number of colonies is increased as much as if the swarm had come off naturally.

By removing the queen.—Some apiarists remove the queen during the honey-harvest and cut out all the queen cells. They give the queen a nucleus if they wish more brood; meanwhile the colony will not swarm without her. Whether queenless bees are as easy in their minds and, therefore, as ready and enthusiastic in the task of gathering honey, is a mooted question. Bees, like people, work to the best advantage when they have fewest worries. One difficulty with this method is that before we are aware of it a queen may be reared despite our careful and onorous labours in hunting for queen cells. Another difficulty with this practice is the encouragement of the egg-laying workers, which is a most demoralising influence to introduce into a hive.

AFTER-SWARMS

An after-swarm is one that is led by a virgin queen and may come off within sixteen days after the first natural swarm departs; usually it occurs within a week. Most bee-keepers consider the after-swarm as a manifestation of "pure cussedness" on the part of a colony; but it is only a poorly adjusted method practised by the bees for getting rid of superfluous princesses. After the old queen decamps with her followers, there are usually several queens ready to emerge from their cells; the ordinary story, as told in books, is that the first queen that emerges hastens to slay her yet helpless sisters, or battles with them singly until but one queen is left in the hive; and this actually does occur often. However, there are differences between bee-colonies, as there are between individual people, and every young queen does not seem blood-thirsty; or perhaps in some instances the citizens restrain her from carrying out her murderous intent, and try to get rid of her by sending her forth with as many followers as can well be spared from the parent colony. Other queens may issue, and if this charitable instinct still persists, another and still another swarm may be sent out. This misguided kindness to young queens is as demoralising to the colony as unwise giving of alms is in the human world, and finally a swarm may be sent off so small that a teacup would hold it. Of course, this means certain death to all of its members. Finally the limit of endurance is reached, and with the last possible swarm are sent out all the young queens left in the hive save the one retained as queen mother. Whether the workers send out these burdensome members of royalty as a measure of good riddance, or whether in their excitement they fail to guard the queen cells and so let them out and they voluntarily join the procession, is not as yet surely ascertained. One of the formalities of the after-swarm is that before it occurs the queen sounds her pibroch, a tune which probably excites her listening subjects to rash departure. The bee-keeper gets to know this sound very well, and when he hears it, he knows that an after-swarm will issue very soon unless he does something immediately to prevent it. One thing particularly exasperating about the after-swarm is that the virgin queen, being lighter bodied and lighter minded than the old queen, may take the occasion of swarming to get married, and go on her wedding journey; and thus is likely to lead her followers a mad chase and leave her proprietor so far in the rear that he loses the swarm entirely.

For the real reason of after-swarming we must look upon the colony as an individual; and as nature is wasteful in the production of individuals, these weak after-swarms are analogous to the weaklings among animals or plants, which must be sacrificed for some inscrutable reason on the altar of the preservation of the species.

PREVENTION OF AFTER-SWARMS

Mr. Hutchinson practises the following method: When the first swarm comes off he places it in a new hive like the old one, and puts the new hive on the exact site of the old one, while the latter is moved away just a little and faces in a different direction than before. The new hive has four or five frames with foundation starters, and on it is placed the super with partially filled sections from the old hive, with a queen-excluding board between the two; thus the new swarm, having no brood ready, will store in the supers until the brood-comb is built. Most of the bees from the old hive, returning from the field, will enter the new hive because the entrance to the old hive is turned away. The old hive is then turned a little each day, until its entrance, which should be contracted, is parallel to the entrance of the new hive, and very close to it. If, on the seventh day after the first swarm issued, the old hive be removed to some new location, its numbers will have been so depleted that there will be no attempt to send off after-swarms.

HOW TO UNITE COLONIES

If two colonies are weak late in the season, it is often best to unite them. This may be done by moving the hives nearer to each other, a little each day, until they are side by side. The queen of one colony is killed, the best frames from each hive are alternated with each other in one hive, and the bees shaken into this; the other hive removed, and the one remaining placed midway between where the two stood. Sweetened water flavoured with peppermint may be sprayed over both colonies just before uniting, so that they will all be scented alike. The above is the method advised by Professor Cook.

CHAPTER VIII

THE HIVE, AND HOW TO HANDLE IT

THE BROOD-CHAMBER

THE essential parts of a hive are the following: A bottom-board, the first story, or brood-chamber, containing frames for the support of the combs, and a cover. When extracted honey is to be produced, a second story like the first may be placed between it and the cover; when it is desired to have the surplus honey stored in section-boxes, one or more shallow stories containing the section-boxes are placed above the brood-chamber; these shallow stories are known as the supers.

Formerly the brood-chamber was a mere cubiform box with two horizontal cross-pieces passing through the centre at right angles to each other for the support of the combs. Sometimes the bee-keeper furnished the bees with a hive made from a section of a hollow log, with a board nailed over one end for a cover, or the hollow log was placed in a horizontal position. Such a hive was known as a "bee-gum," probably because it was often made from the trunk of a gum tree; but the bee-gums with which we were familiar in our childhood were made from hollow basswood logs.

To-day in every well-regulated apiary the brood-chamber is furnished with the movable hanging frames for the support of the combs, which were invented by Langstroth a half-century ago, or by some modification of these frames. It was the invention of these frames that made the science of modern bee-keeping possible. A large part of the manipulations of the hive is dependent upon the ability to remove the combs from the hive separately. Two of these frames, one empty and one containing a sheet of foundation, are shown in Plate XII. These are of one of the newer styles, known as the Hoffman self-spacing frames. In these frames the upper part of the end-bars are wide, so that, when the frames are in contact, there is room for a bee-way between the combs; the lower part of the end-bars are narrower, so that the bees can pass freely around the ends of the frames.

In the old days, and at present in some apiaries where home-made frames are used, the spacing between the frames has to be done by eye or rather by finger; the thickness of the tip of the finger between two frames being necessary to afford a bee-way. But with the new frames in the market to-day this is done away with, as they are arranged to space themselves, thus relieving the novice of much anxiety and some embarrassment in deciding whether his finger is as thick as that of the average apiarist. Also in these newer frames a little staple at each end of the top sets each frame exactly right in relation to the ends of the hive.

Before the frame is placed in the brood-chamber it is to be supplied with a sheet of comb-foundation; this insures the building of the comb in the desired position, so the frames can be removed from the hive. Bees do not naturally make their combs straight, but curve them so that they are less likely to be broken. It is necessary, therefore, in order to get straight combs to confine the operations of the bees within set bounds; this is done by putting a sheet of foundation in each frame.

One of the attractions of bee-keeping is that much of the material we work with is pleasing, and some artistically beautiful. A sheet of foundation-comb made of the most delicate wax, frescoed on either side with rhomb insets in hexagonal pattern, is a joy to the artistic eye. It is absolutely necessary that the foundation should be put into the frame in such a manner that it will not sag or bulge when it is built out into the comb and filled with the heavy honey or brood. To keep these large sheets from bulging and breaking they are held in place by fine wire which is strung back and forth across the frame, passing through holes made with a small awl in the end-bars. These holes should be about two inches apart, the upper one one inch from the top bar, and the lower one something less than an inch from the bottom bar, making four wires in the Langstroth frame. After these holes are made, put a small tack at one end near the lower hole, twist the wire around it, then thread the wire back and forth, making four wires parallel with the top and bottom bars, and fasten the wire with a tack; care should be taken not to draw the wire too tightly; simply draw up the slack.

The sheets of foundation as sold by the dealers are a little smaller than the space in the frame, so that when they are fastened to the top bar a bee-space is left between the sheet and the bottom and end bars. See Plate XII.

The sheet of foundation is fastened to the top bar of an ordinary frame by means of a Van Deusen wax-tube fastener, which is simply a hollow tube that may be dipped in and filled with the hot wax which issues through a small hole at the sharp, bent end of the tube; as the point is drawn along where the foundation and frame meet it leaves a stream of hot wax that seals the two together. However, the most satisfactory way is to get brood-frames, like the Hoffman, which have two grooves in the top bar. Set the foundation in the groove at the centre, and introduce a strip of wood which is wedge-shaped in cross-section, thin edge first, into the adjoining groove, driving it or pressing it in hard so that it pushes the thin partition over, and wedges the foundation firmly in place. These strips for wedging come with the frames.

After the foundation is fastened to the top bar the frame and foundation are laid, wire side up, on a board just the size of the piece of foundation so that it will slip inside the frame. This board is kept wet to prevent the wax from sticking to it. Then we use the spur wire-embedder, which is like the tracing-wheel used by dressmakers, except that the teeth are spread apart alternately so that they pass astride the wire and press it down into the foundation; to do this successfully, the foundation should be warm; working near a lamp or in a warm room will suffice. Embedding the wires by heating them with electricity instead of using the spur-embedder is a common practice in large apiaries where electricity is available, or where it pays to buy a battery with proper attachments. This outfit with the dry cells costs about five dollars, and is a paying investment when the apiary is large. (Plate III.)

THE SUPER

The super is that part of the hive that is placed above the brood-chamber and is designed to receive the surplus honey, either comb or extracted. When extracted honey is produced the super may resemble the brood-chamber described above, or it may not be so high, fitted to receive a frame little more than half as deep as the standard Langstroth frame used in the brood-chamber.

When comb-honey is produced, the super is only about half as high as the brood-chamber. For this reason a hive consisting of a brood-chamber and one super for comb-honey is termed a one-and-a-half-story hive. (Plate XV.)

A complete super fitted for comb-honey consists of the following parts: (1) The outer wall, which is of the same length and width as the brood-chamber and of the right height to hold the style of section-boxes used; on the lower side of each end a tin strip is fastened to support the fixtures that it is to contain; (2) the section-boxes; these are the wooden frames containing the comb-honey when it is placed on the market; each holds about one pound of honey; four of these are shown in the Plate XIV; (3) the section-holder; this is a rack fitted to hold and support one row of section-boxes, as shown in Plate XIV; when in place in the super, it rests on the strips of tin mentioned above; (4) the fence; this is a device placed between the rows of section-boxes to keep the bees from building the comb beyond the edges of the section-boxes; the style used with the plain or no-beeway section is shown in Plate III; the vertical cleats on this fence provide for a bee-space between it and the section-boxes, so that the bees can build out the comb even with the edge of the section-box; (5) the super springs; these are three flat springs, fastened to the inner face of one side-wall in such a way that they press the fences and the section-boxes closely together. (Plate XII.)

There are several types of supers in use that differ, in certain details, from the one described above. In some, the rows of section-boxes and fences are pressed together by thumb-screws which pass through one side-wall of the super. Many bee-keepers still use section-boxes with beeways, that is, boxes having the top and bottom narrower than the two sides. When such boxes are used, the fences lack the vertical slats, or posts; in fact, a simple strip of tin may serve as a fence.

Most of the section-boxes in use now are made of a single piece, which is dovetailed at the ends and has three transverse V-shaped grooves cut in one side so that it can be bent into shape, as shown in Plate II. These flat basswood sections afford very pretty material with which to work. The novice, in putting them together, almost always bends them with grooves outside at the corners, instead of inside, and then wonders why they are askew, and break. We wet the sections where they are grooved before we begin working with them. This may be done by brushing each individual flat with warm water, which is a very tedious process; but it is better to take a handful of them as they lie, the grooves all in a line, set them on edge and pour a little water from a pitcher on the grooves. This wets many at a time with little trouble. To set up a section, it should be taken in the hands, grooves up, bend the ends upward evenly, fastening the dovetailed edges together gently. Haste and jerkiness are as disastrous in handling sections as in handling bees.

The foundation for the comb-sections is much lighter in weight than that intended for the brood-frames. Some apiarists fill the entire section with this foundation, except for the bee-spaces at the bottom and at the sides. But we never do this, unless we are obliged to do it to coax the bees to use the supers; for it is not so satisfactory as to use a narrow strip for a guide at the top of the section, as a starter to show the bees where to build the comb. We cut these strips about an inch deep and almost as wide as the section, and, with the Daisy fastener, fix each strip at just the middle line of the upper bar of the section. The objection to filling a section with a complete sheet is, first, the expense of the foundation; and, secondly, that it is likely to give a tough central portion or "fish-bone" to the comb-honey. (Plate XIV.)

HOW TO PREPARE THE SUPER

If they are not ordered set up ready for use, the supers come in flat pieces with dovetailed ends, and putting them together is a pleasing occupation, after one has learned how. The best way to learn how is to carefully observe a super already properly set up; for, though the directions for putting these together are as plain as may be, yet a person may err therein and yet not be a fool. Unless one has learned, or can learn, to drive a nail, one had best not undertake bee-keeping, for the bee-keeper must become a carpenter to be successful; it adds much to the interest of the occupation to make all sorts of things for one's own bees. The principle on which the super is built is that it may hold the sections tightly in place, and not allow them to drop through. Therefore, at the bottom of the super, along each narrower end, is a tin strip to support the ends of the section-holders; to keep the ends of the section-holders even a wedge-shaped strip of wood is nailed across the end of the super, thick edge down and flush with the bottom edge, resting against the tin strips. We use the Hetherington super-springs, one at each end and one at the middle of one side of the super; the trick of putting these in is to set

(a) Two self-spacing frames; one of them fitted with a sheet of foundation.

(*b*) An empty super, showing the springs and the wedge-shaped piece at the end.

(*c*) The Doolittle division-board feeder.

PLATE XII

Photograph by Brown Brothers

PLATE XIII. EXAMINING THE BROOD-FRAMES

them opposite the posts of the fence, one at either endpost, and one at the middle post, else the spring will be of no use. Driving in the sharp end of this spring successfully requires a little practice. The super is now ready to be filled. (Plate XII.)

First place a fence in the super on the side opposite the spring; then a section-holder filled with four starters, always remembering that these foundation-starters should be at the top of the section. Then place another fence and another section-holder. The eight-frame super will hold six of these rows, with fences between and one on each side. Putting in the last fence is the final test of whether the springs are in the right place, for, if they are right and press against the posts of the fence, it will be very hard to push in this last fence;

when it is in, the sections are all held snugly in place. (Plates III, XV.)

Where the super is placed on the hive, it should closely fit the top of the brood-chamber, with no cracks between. If the hive has a flat cover, which leaves only a bee-space above the sections, the cover may be placed immediately above the super, with nothing between. With covers like the telescope cover, a super-cover is needed. This may be a quilt or a piece of enamelled cloth; but we prefer a super-cover made of a thin board, bound on the ends to prevent warping, which is now on the market.

HOW TO HANDLE THE BEES

It is generally believed, and for good reasons, apparently, that bees like some people and despise others, who are just as good, so far as we can detect. This apparent capriciousness has been explained in many ways. Some hold that the bees have a fine sense of smell, and thus distinguish us by odours rather than by sight; and in this case their ire is aroused because they do not like the perfume exhaled by the obnoxious person. Others claim that it depends upon the movements; if one moves nervously and quickly, he is much more subject to attacks. It is certainly true that if a bee, which is buzzing threateningly, is struck at, she becomes more enraged and is more certain to sting, but this is because she recognises an aggressive foe because of the act. However, the senior partner in our apiary is an exceedingly active and nervous man, and I have seen him move with all haste and energy while working with bees, and though he seldom uses bee-veil or gloves, he is rarely stung. Our bees seem to be acquainted with him, and accept his rapid movements as one of the common-places of bee existence. It is well for anyone who wishes to work with bees to spend some time in the bee-yard just watching the little citizens coming and going, and listening to the peculiarly soothing hum which always fills the air around the hives. It is sympathy with the bees more than actions that finally results in handling them without harm.

HOW TO OPEN THE HIVE

First of all, fire up the smoker. The way to do this properly is to place some easily ignited material in the bottom of the fire-chamber, touch a match to it, and crowd in above it material which will make plenty of smoke, and will not burn too rapidly; give a puff or two with the bellows to be sure that the fire is started. We have used excelsior in our smoker because it was near at hand, but it is not a perfect or lasting fuel. Fine chips, especially planer shavings, old rags, greasy cotton waste, and even pine needles are used. Anything is desirable that will make a smudge and will not burn out too quickly; for when we are working with the bees we have little time or inclination to stop and "putter" with the smoker; and we cannot afford to have the smoke give out at the critical moment when we most need its protecting incense. A minimum amount of fire with the maximum amount of smoke is the desirable quality in the smoker. If the fire gets too hot the blasts will burn the bees, which is an outrage, and which is never permitted by a civilised individual. The Cornell smoker has a hook attached to the bellows, so it may be hung on the edge of the open hive to be at hand in time of need. If it becomes too hot, we lay it flat on the ground so as to stop the draft. Each time after the smoker is used it should be emptied, otherwise it is likely to start a conflagration. We have an ash pail near the apiary in which we always empty the smoker on our way back to the bee-room. Mr. Root speaks of never using the smoker until it is needed; when his bees trouble him, he gently pats them on the back with a little grass to get them out of his way as he lifts up the frames. And we never can admire Mr. Root enough for dealing thus gently with his beepeople. But we would not advise the novice to try this, as a person has to be on very intimate terms with bees to be able to pat them on the back with grass and impunity. However, this is an ideal to work toward. The nervous beginner almost invariably uses too much smoke, and this makes his little dependents unhappy. The breathing of smoke is hardly a pleasant experience for us, and it seems to be still more distressing to the bees. We remember once how, in the enthusiasm of our novitiate, we inadvertantly smoked the bee-man instead of the bees in our misguided efforts to help, and the result was a blueness of the atmosphere which rendered more smoke superfluous. Every beginner ought to get at least one headache from the fumes of the smoker to teach him charity and care.

There are several reasons why the hive must be opened, aside from the fun one derives from the experience. First, the brood needs to be examined occasionally to see that it is all right, and in the fall the brood-comb must be examined to see that there is enough honey stored within it to winter the bees. Second, during the swarming season to find and remove the queen-cells. Third, to hunt for the queen to be sure she is present and active, or perchance to find her and clip her wings. Fourth, to take off supers filled with honey. A warm day should be selected for opening the hive for whatever reason, and the middle of the day is the best time for the work.

Send two or three puffs of smoke in at the entrance to drive back the frightened sentinels who keep careful watch of the portals of the hive. Then lift one edge of the cover of the hive a little and send two or three puffs in the crack; then lift off the cover and set it down beside you; then lift the quilt or super-cover at one edge, and give two or three puffs of smoke beneath it to drive the bees down among the frames, always remembering that under ordinary circumstances a very little smoke is necessary to frighten and subdue the bees. In fact the same rule applies to smoking bees as to smoking tobacco; if one is moderate in its use, the least harm will result.

HOW TO EXAMINE THE BROOD-CHAMBER. (Plate VIII.)

Stand at one side of the hive, and not in front of it. Hang the smoker on the side of the hive, so as to have it within reach. Mr. Root advises sitting on the cover of the hive set on edge while you examine the brood-frame. This will do for a well-poised person, but we prefer a little stool, which we can carry easily from hive to hive, as we wish something that we can sit on calmly as the situation requires. Commence at one side and loosen the outside section with a knife, or what is better, an old screw-driver. Take the frame by the projecting ends, and lift it up so that you may examine it on one side, then twirl it half-way over to examine it on the other. It requires some experience to placidly lift out a frame, covered with what at first sight looks like a dark, boiling, viscous fluid, fit only for a witch's cauldron, but which soon to the startled eye resolves itself into bee particles. If the brood must be examined or queen cells found, then it becomes necessary to get rid of this seething, enveloping bee-mass, which is done in a manner that seems like nothing less than tempting Providence. The brood-frame is seized firmly in the operator's hands, and held about waist high in front of the hive, then let to drop, hands and all, swiftly to within about six inches of the doorstep of the hive, then suddenly jerked back again; the bees being heavy and receiving the downward impetus, keep right on as a man keeps on when his bicycle stops suddenly in a rut—with this difference that the bees land safely at the entrance of the hive, into which they scamper as soon as their dazed wits will allow. One would naturally think that the bees would attack the active agent of this indignity, but while bees are ever ready to fight recognised enemies, they have evolved no plan of action which is equal to a cataclysm, except to get under cover as soon as possible. Thus being shaken from their foundations is to them what an earthquake is to us, and the attitude of Riley's boy is theirs.

"Where's a boy a-goin' an' what's he goin' to do,
An' how's he goin' to do it, when the world bu'sts through?"

Thus it is that among the first accomplishments of the apiarist is the one that enables him skilfully and with dispatch, and without harming the individuals to shake the surprised little mob off the frame or out of the boxes, to a place where its members may recover shelter and equanimity as expeditiously as possible.

After the comb is freed from the bees, it requires some experience with honey-comb topography to see at a glance just the condition of the brood. There may be cells that look empty until a ray of light reveals at the bottom a glistening egg; and there may be cells with a little milky substance at the bottom in which the young larva is floating; or in some cells the bee grubs may be distinctly seen if they are four or five days old. If the cells are capped, it may puzzle the novice to know what lies behind that closed waxen door. If the cells contain honey, the substance of which the cap is made is whiter than that which covers the brood. In case of worker-brood the cap is depressed slightly below the plane of the comb, which is not the case if the cells contain honey. The large size of the drone cells distinguishes them readily from the cells of the workers. Often honey is stored in drone cells, for the bees seem to like to make these larger cells, and for good reason, since they give greater storage capacity for the amount of wax used. However, the drone cells which contain brood are covered with dark, dirty, yellow caps which are quite convex, looking like kopjes on the comb plain. At the height of the honey season there should be plenty of brood, and later the cells in the brood-frames should be filled with honey. The cells containing bee-bread are not capped, as this staff of bee life is packed so hard that it does not need to be covered. All honey remains uncapped until it is properly evaporated and ripened. (Plate VIII.)

After one frame has been thus examined, it should be leaned up against the side of the hive so as to give space to lift out the next frame without crushing the bees.

HOW TO FIND THE QUEEN CELL

Fortunately for us, this is quite prominent, being a veritable oriel in shape. However, there may be other excrescences of the comb which somewhat resemble a queen cell; sometimes the queen cell may be more or less embedded and so escape observation. The bee-keeper who is cocksure that he can find all the queen cells in his hives has to be most experienced, and even then cocksureness may come to grief. But this unglazed oriel window in which the queen develops is usually quite noticeable, and is ordinarily decorated with a small, hexagonal pattern in relief. We have often wondered if this was done for the sake of decoration, or because the bees are so in the habit of fashioning wax into hexagonal patterns that they do it involuntarily. For the person who rashly asserts that honeycomb is the result of fortuitous force and pressure, this queen cell with its hexagonal frescoes is a poser. (Plate V.)

To cut out the queen cell a sharp, pointed knife is necessary, in order to injure the comb beneath the cell as little as possible.

HOW TO FIND THE QUEEN

If we simply need assurance that the queen is present and active, the discovery of eggs or young larvae in the cells is sufficient evidence of her presence, and saves the tiresome search for her majesty. But if we wish to find her, she is usually present on the middle frame of the hive. It is not safe to pull out this middle frame from the narrow place which it occupies for fear of hurting the queen and crushing the other bees; so it is best always to take out an end frame, first looking at it carefully to make sure she is not upon it; then shake off the bees and set it beside the hive, and move the other frames along in the space thus made until we are able to remove the middle frame with ease. It requires some experience to ferret out the queen from the bee-mob which seethes over the comb. The burly, big, bluntended drones are much more readily detected. However, after a little training in the devious ways of royalty one becomes expert in seeing that long, graceful, pointed body extending far back of the wings which characterises the queen. If the bees are not too much disturbed, she is likely to be surrounded by a rosette of workers, all with their faces toward her, for even in the court of the hive etiquette does not permit that the ladies-in-waiting turn their backs to the queen. If for any reason the queen is to be lifted out, she should be seized by the wings or thorax or imprisoned in a queen trap, but never under any circumstances should she be seized by the abdomen. (Plate XVI, Queen trap.)

CLIPPING THE QUEEN'S WINGS

This does not mean cutting off all four of the wings, but that the wings on one side should be clipped, leaving stubs not more than an eighth of an inch long. Various devices have been invented to aid clumsy hands in cutting off the royal petticoats. One, the Monette queen-clipping device, is a little cone-shaped cage made of wire laid in spirals. She goes into this cage head first and the door is shut behind her. Then the scissors are slipped between the spirals at the proper point and the deed is done. Another simple device is a bit of section-board whittled in the shape of a tiny bootjack with a rubber band stretched rather tightly across the prongs. The forks are placed across the queen so that the rubber presses against the thorax, thus pinning her fast to the comb while she is barbered. The only skill needed in this device is in fixing the tension of the rubber band so that it will be sufficient to hold her majesty fast, and yet not stiff enough to hurt her. Our invariable plan is as follows: After the queen is discovered, we hold the brood frame in one hand, pick up her royal highness in the other most gently, then still more gently set the frame down, leaning it against the hive; then, holding her royal person firmly but carefully in our own unworthy thumb and fingers, clip her wings with presuming scissors; then, putting the scissors down as we pick up the frame, and put her back as nearly as possible on the spot where we found her. We always use the sharppointed embroidery scissors for this delicate operation.

HOW TO TAKE OFF HONEY IN SUPERS

Lift the hive cover and quilt with a slight introduction of smoke, then lift off completely. Smoke from above for a moment, being very careful not to burn the bees, always remembering that smoke is meant to scare and not to punish. Then loosen the super with a screw-driver if the bees have fastened it down with bee glue, lift it and place it on a bottom-board near at hand. Put on an unfilled super if it is needed and cover the hive. Lift the honey-sections out of the super, brushing off the adhering bees with a bee-brush, so that they will fall on the doorstep of the hive, and place each section as it is cleaned in a basket or box in which it is to be carried to the store-room. This is our usual way of procedure when our apiary is small, and we do this work during spare moments which are not predictable the night before. However, there is one best way to do this work, and that is to put on a Porter bee-escape the night before the super is to be removed. Wise from experience, we advise beginners to study this device and become imbued with a knowledge of its workings to the extent of being able to tell which is the upper and which the under side, lest disastrous results ensue and all of the bees escape into the super instead of out of it. If the bee-escape is placed on the night before, there will be no bees in the super when it is removed next day. To introduce a bee-escape one does not need to lift off the super; simply lift it up at one side, send a little smoke into the crack, push in the bee-escape, and then set it straight upon the hive and the super straight upon it. (Plates III, XVIII.)

The novice might conclude that a good plan would be simply to set off the super near the hive, and let the bees find their own way back to their brood and kindred. But bee-nature has to be reckoned with in this instance, and the bees, conscious that the honey is their own, are likely to uncap the cells and carry the honey into the hive; or worse still, the bees from other hives will be attracted to these open stores and will begin to rob. And in the bee courts of equity, when bees begin to rob, then the "devil is to pay."

There have been various bee-tents devised under which the supers are placed after being removed from the hive. These tents are arranged with a little hole at the top by which the bees may escape, but may not return. Doctor Miller invented a simple plan of piling several supers filled with honey on top of each other, leaving no crevices between them; over these was spread a cloth with a hole in the middle, over which was placed a wire cone with a hole in the top large enough to let a bee pass out. Thus the bees from all of the sections escaped, one by one, and robbing was avoided. However, the bee-escapes introduced between the super and the hives are used to-day by most enterprising bee-men.

OBSERVATION-HIVES (Frontispiece)

Anyone who has worked long with bees, cannot fail to become filled with curiosity concerning the way their work is carried on in the mysterious darkness of the hive; to such a person, the observation-hive is a source of perennial delight, as well as of interesting and useful knowledge.

Observation-hives have been used by bee-keepers from the time of Huber to the present, and naturally many forms of them have been devised. The type in most common use now is a small hive, containing one, two or three frames, and furnished with glass sides, through which the bees can be observed. The glass sides are covered with a door or curtain, except when observations are being made; for, if not, the bees will cover the glass with a coat of propolis, rendering it opaque.

It is somewhat difficult to keep a colony in good condition upon a single frame; and if two or more frames are used side by side the observer is unable to see what goes on between the frames. Professor Kellogg, of Stanford University, has devised a perfectly satisfactory two-frame observation-hive for his laboratory. It consists of a glass-sided box, large enough to hold two Langstroth frames, one above the other; as both sides of the comb are exposed, any individual bee may be kept constantly in sight while she is working. The passage which leads from the hive to the opening in the window has a glass top, so the actions of the bees as they enter the hive may be watched. A hood of thick black cloth is pulled down over the hive when the observations are finished. Similar observation hives may be purchased of dealers in apiarists' supplies. Such an observation-hive would be of great value to the enterprising bee-keeper, as it would be the means of helping him to understand conditions which were puzzling, and thus aid him in dealing with crises that are sure to occur. The advantage of this hive over any other is that frames from any hive may be kept under closest observation. But if the hive were a means of interest merely, it would still be worth while, for the bee-keeper cannot know too much about the ways of bees, *supposing all he knows is true*.

MAXIMS FOR OPENING THE HIVE

Have the smoker ready to give forth a good volume of smoke.

Use the smoker to scare the bees rather than to punish them.

Do not stand in front of the hive lest the bees passing out and in take umbrage.

Be careful not to drop any implements with which you are working; take hold of all things firmly.

Move steadily, and not nervously.

Do not run if frightened, for the bees understand what running away means as well as you do.

If the bees attack you, move slowly away, smoking them off as you go.

If a bee annoys you by her threatening attitude for some time, kill her ruthlessly.

If stung by a bee, rub off the sting, instead of pulling it out with the nails of the thumb and fore-finger and thus forcing more venom into the wound.

Ammonia applied to the wound made by a bee-sting will usually afford immediate relief.

CHAPTER IX

DETAILS CONCERNING HONEY

HOW TO MAKE COMB-HONEY

EVERY bee-keeper who sends to the market honey in the comb enfolded in an attractive carton, or with the section neatly glazed, has produced a work of art; for comb-honey as now marketed is an æsthetic production, and the bee-keeper is an artist as much as if he had painted a picture or had fashioned a jewel. To most people who have an apiary as a pleasurable adjunct to life in the country, the production of comb-honey is most attractive, while the production of extracted honey does not appeal to them at all. Just the word "honey" calls to the mind of most people a vision of amber sweetness set in white-walled, waxen cells.

The production of comb-honey is attended by more difficulties than is the production of extracted honey. The reason for this is largely because the bees work more readily in cells already made from which the honey has been extracted, than they do in sections where they must undertake all of the expense and labour of producing wax for the comb. More than this, honey may be extracted from the comb

(a) Alley's queen and drone trap.

(b) A well-filled section

(c) One empty section-holder, and one filled with section-boxes, in which are foundation-starters; two of these have been added to by the bees.

PLATE XIV

Photograph by Brown Brothers

PLATE XV. One and a half story hive for comb-honey; the super is filled with section-boxes.

of the brood-chambers, a harvest which is lost to the producer of comb-honey.

For the production of comb-honey it is necessary that the colony winter in excellent condition and develop much well-fed brood early in the season, so that there shall be a great number of active young workers in the hive just before the chief honey-harvest of the summer begins. In New York State we have two large honey harvests, that of the basswood in July and the buckwheat in August, so our colonies are made strong and ready by the last of June. In order to have the bees ready to work, the swarming fever must be subdued or controlled before this period of honey-flow. The colonies are carefully watched early in the season, and if after the first pollen-gathering occurs there is no honey coming in, the bees are fed so that the brood may be developed. We rarely have to feed at this season of the year as the fruit bloom gives our bees plenty of honey for rearing their brood, and we never expect any surplus before the basswood season.

The supers are put on when there is plenty of brood and plenty of honey to feed to it in the hive; and under such conditions our bees, though Italians, usually push up into the sections at once. We prepare the sections with mere strips of wax foundation for starters. (Plate XIV.)

Supers thus equipped put on at this time and under such conditions seem to take away all desire to swarm on the part of the bees, if the queen cells have been previously removed.

If the brood-chambers are crowded when the super is put on, the queen may go up into it and start brood. This rarely occurs with us, but in case it does a queen-excluding honey-board may be introduced between the super and the brood-chamber. Sometimes a colony seems unwilling to go into the super after it has been added, and the bees will hang on the outside of the hive and threaten to swarm, and through doubt and vacillation lose two or three of the precious days of the basswood harvest.

The usual reason why the bees will not go into the super is that they are not sufficiently crowded below; if they have room they prefer to store their honey in the brood cells rather than carry it to the upper story.

If a colony is strong and has plenty of brood and honey and is still obstinate, men of experience advise the taking of a frame of sections from a colony which is storing in the supers and putting bees and all in the midst of the unused super of the reluctant colony. We have never tried this because we were never obliged to. It sounds very practical and sensible, and it is practised by Mr. Root, and that is sufficient recommendation.

Our way of coaxing the bees into a super which they have sedulously ignored is to place in it some of the imperfect sections, which are not worth much in the market and which contain some capped honey and many empty cells. One section-holder filled in this way seems to encourage the reluctant colony to climb and to store as rapidly as possible.

In large apiaries the following plan is followed: First, put on a super of shallow extracting frames from which the honey has been removed; as these cells are all ready, the bees are likely to go to work in them at once; and after they are working well raise the super, and put in one below it filled with sections containing starters, and all will be well. In dealing with this phase of bee-keeping it is well to remember that full sheets of foundation in the sections are more attractive to the bees than starters; and that sections containing comb already made are still more pleasing, and if some of these made cells contain honey, their attractiveness is doubled. One condition should be observed in putting on a super in the summer: it should be shaded in some way; if it is in the direct rays of the sun, the heat is likely to keep the bees out of it.

However, later in the season bees may be induced to work in the super by placing over it a cushion so that it will be warm.

TIERING UP

If the honey is coming in at a good rate, and the bees are working well, when the sections in the super are something more than half full, lift it up and place another containing sections with starters beneath it. The reason for this is that the bees would not naturally go into the empty super if it were placed on top until the other was completely filled. But with this plan they continue working in the super, even though it be on top, and meanwhile find it "handy" to fill the intervening sections. If the honey-flow is great, even another super may be placed next the hive and below the other two. However, in our practice we rarely put on more than two, usually taking off the top one when we need to interpolate another. This process is called "storifying" in the English books, which is a most graphic term and should be introduced into our nomenclature.

TAKING OFF SECTIONS. (Plate XX.)

In taking off the sections we do not need to wait for the completion of every one in the super. The outside rows are rarely perfect, and we usually put these unfinished sections back on some other hive to be finished.

These unfinished sections, if not too empty, serve very well to sweeten the daily bread of the home table. If sent to the market they bring low prices, and the bee-keeper who is working for comb-honey should plan to have as few of them as possible.

While the honey left long in the super has a much finer flavour than that which is removed early, yet care should be taken not to leave the sections on the hive so long that the comb becomes soiled. It is an interesting fact that honey ripened in the hive gains special richness, as if it were somehow imbued with the spirit of the little socialists that make it.

Toward the end of the season it is best not to tier up, but to place an empty super on top. The bees will not use it unless necessary, but will devote their energies to the sections below. The great danger

Photograph by Brown Brothers

PLATE XVI.Drone and queen trap, below; queen mailing and introducing cage, above at the left; queen-protector and queen-cage used in queen-rearing, in centre; and bee escape, at the right.

Photograph by Verne Morton

PLATE XVII. "IN APPLE-BLOSSOM TIME"

to be avoided in tiering up is a surplus of partly filled sections, and the way we avoid this is not to interpolate a super until the one on the hive is at least three-fourths filled.

STAINED SECTIONS

A regular part of bee exercise consists of promenading up and down and across the sealed honey; the bee has not as yet, unfortunately, attained the fastidiousness which leads her to wipe any of her six feet carefully before entering her domicile, consequently the sections of honey thus walked over may be stained and unmarketable. There is no remedy for this except to look after the supers carefully, and take out the sections before they are soiled.

Some sections may look dirty because old wax is used in making the caps. If such is the case, and it is simply yellow, the wax may be bleached by standing it in the sun or by subjecting it to sulphur fumes. Some apiarists have special rooms and others tight boxes for the sulphur bleaching. Only two things are necessary to accomplish this successfully; first, that the room or box be tight; second, that the sulphur placed in an iron dish be heated so that the fumes are strong and all-pervading. Some say that the sulphur should be heated so that if a match be touched to it it will flame. The combs need not be subjected to such fumes more than a half-hour to become as white as they can be bleached.

SCRAPING THE SECTIONS

In preparing comb-honey for the market, it is necessary to first scrape the propolis from the sections so as to leave the wood white and beautiful. To do this the section should be set square on and in line with the edge of a table, and below should be a pan to receive the scrapings. Hold the section firmly in one hand and scrape the side that is in line with the edge of the table with a downward stroke of the knife; a case-knife is best for this. To scrape the wood clean and not in any way injure the honey, and to do the work rapidly, measure the skill attained in this business. In scraping the sections it is best to have four of the shipping-boxes at hand, so that the honey may be graded and placed in its proper class as it is cleaned.

GRADING COMB-HONEY

First, as to the way the sections are filled, or in other words, as to the bee technique. There are three grades—Fancy, No. 1, and No. 2. In the fancy grade almost every cell is well filled, and the comb has the surface evenly built and well capped. No. 1 has an even surface and is well filled, but may not be so perfect in the corners as is the fancy. No. 2 must be at least three-fourths full. Anything below No. 2 is called chunk honey if sold in the comb, but it is more profitable to extract honey from all sections that range below No. 2.

Second, as to the colour of honey: It is graded as white, amber, buckwheat and dark, and these need no explanation. Thus honey is listed perhaps as "fancy white," or "fancy buckwheat"; in each case it is the best of its kind.

SHIPPING-CASES

These should be ordered rather than made, as a good-looking shipping-case adds materially to the value of the honey. These cases come in different sizes, and have one side made of glass so that the handlers may see that the contents are fragile, and therefore may possibly be persuaded to deal with them gently; the cases come in flats and are easily put together. For anything that looks so well put up as honey sealed in its perfect cells, it has a most amazing capacity for leaking. Once, in the enthusiasm of girlhood and inexperience, I carried some honey carefully packed in a box in my trunk, hoping to give a friend a treat. Needless to say that honey was "linked sweetness long drawn out," by the time I arrived at my destination; and all the clothes that I carried in my trunk were literally "too sweet for any use."

Shipping comb-honey to market is likely to be a disastrous performance at best, since it is almost impossible to guard against careless handling. Some ship in glass-covered sections which protect the comb, and make a very attractive appearing product.

Shipping the sections in cartons is winning its way now for fancy grades. A carton is a pasteboard box which may be bought in flats with the shipper's name printed upon it. While these cartons may be bought in almost any style, from perfectly plain to those highly ornamented, and provided with tape handles, yet we believe there is a chance for personal initiative in this particular field. An artistic design in pretty colours, individual and unique, would certainly prove a special attraction for selling honey in cartons. The great advantage gained from the use of the carton is that the honey reaches the consumer in a neat package without further handling, and may be carried like a box of bonbons. When cartons are used the shipping case should be a size larger than for the plain sections.

There are two ways of marketing honey open to most bee-keepers. First and best, the local market. If the honey can be placed in the hands of the grocer directly from the bee-keeper, certain advantages accrue. The comb is not broken by much careless handling, and it reaches the market in good shape. Though the price may be somewhat lower perhaps than the highest quoted prices, yet it is reliable, and there is no discount for breakage in shipping, and for differences of opinion in grading. It is certainly far more satisfactory for all concerned to place comb-honey on the home market; this is usually practicable for all except the greater apiarists. We know one man who has about forty hives, and who lives in a town of about three hundred inhabitants; though he produces a reasonable amount of honey he very inadequately supplies the demand for it in this little village, and he receives city prices.

The second and least desirable method of marketing comb-honey is to ship it to a commission merchant. If this is done, it is well to select a middleman in whom we have absolute confidence, or we are likely soon to become pessimistic regarding his honesty; so frequently is the price of honey reduced on account of breakage and leaking and other accidents which this very frail delicacy is heir to, that we rarely realise the prices quoted in the newspapers. If a good middleman can be found, then our advice is to stick to him, and send him the very best product possible, fairly graded and in the most attractive form, hoping that he may be able to do for us what we should do for ourselves, and that is, work up a special market.

STORING COMB-HONEY

It is far better to market comb-honey the year it is made. However, if it is to be stored, it must be placed in a room that has a constant temperature above 60° F. It is best to fumigate the sections if there is any danger from the bee-moth, for this little rascal will destroy a great amount of comb-honey in a very short time. (See "Bee-Moth.")

CANDIED COMB-HONEY

Some kinds of honey will granulate much sooner than others. The longer the honey is left in the hive and the more perfectly it is ripened, the less liable it is to granulate. Extracted honey will candy much sooner than honey left in the comb. We have kept comb-honey more than a year without crystals appearing in it. The only way to prevent comb-honey from candying is to keep it in a temperature that does not fall below 60 degrees. After honey is candied in the comb, nothing can be done with it except to sell it at a lower price, or keep it to feed back to the bees. The latter is probably the most profitable way to dispose of it. Some people like comb-honey after it is granulated and the home-table may use a certain amount of it.

MAXIMS FOR PRODUCING COMB-HONEY

Keep the colonies strong.

The bees should be kept warm and well fed in the spring.

The bees must have wintered well.

The colony must have brood and plenty of honey in the brood-chambers at the beginning of the honey season.

Never let the honey in sections or supers be exposed in the apiary to incite robbery.

Keep the sections in a room in which the temperature never falls below 60°.

Fumigate the sections before they are stored if you are troubled with bee-moth.

Send the honey to market in as attractive form as possible. Make your product individual in appearance, and strive to create for it a special market.

CHAPTER X

EXTRACTED HONEY

HOW TO PRODUCE IT

It is certainly fortunate for bee-keepers that centrifugal force is one of the unalterable laws of the physical world. However, this force might never have been of any use to the apiarist had it not been for a certain Major Francesco de Hruschka of Venice, who is a most interesting figure in the history of bee-keeping. Next in importance to the invention of the movable frames by the venerable Langstroth was the invention of the honey-extractor. We have a picture of Major Hruschka in our minds as not only a brave man of war as indicated by his military rank, but also as a happy man of peace, who loved his hives with their little citizens, and who also possessed notable domestic virtues and loved the companionship of his children and was interested in their doings. For it was when his little son accompanied him to his apiary and, having a comb full of honey set up in a basket, began whirling it boy-like by the rope attached to the handle, that the Major discovered the honey was being thrown out of the comb by this action. Instead of spanking the boy as most fathers would have done, the thoughtful Major cogitated on the fact that such a simple motion should have emptied the comb of honey, and he straightway proceeded to invent the first honey-extractor. This was in 1865; up to that time the liquid honey was extracted by a method which we remember well. The comb was crushed, and with it too often, alas! the dead bees, larvæ and any dirt whatsoever that happened to be present; this mixture was suspended in a cloth bag, over a tub or vat in a warm room; and the honey, carrying with it much of the débris, slowly dripped out reeking with an aroma and a flavour quite unknown in these regenerate days, and forming a product that may well be spared from the world's marts.

Since 1865, many honey-extractors have been invented in America, and almost all of them which have survived the test of use are satisfactory. The perfection of the invention is an automatically reversible machine with ball bearings, highly geared in order to attain the maximum of steadiness and rapidity with the appliance of a minimum of power.

The principle of construction underlying all of the best extractors is a cylindrical can containing wire pockets in which the combs are set on edge, and which are revolved by being geared to a crank at the top of the can. There is room for more or less honey below the wire pockets, and at the bottom of the can is a faucet or honey gate, through which the extracted honey may be drawn off into a pail or vat.

The extractor is a very excellent adjunct to any apiary, however small, even if comb-honey is the chief product, for it saves much honey that otherwise would be wasted. When the apiary consists of less than forty hives of bees, one of the small non-reversible extractors may be used. These weigh less and cost less; but every frame of comb has to be taken out after the honey has been extracted on one side and reversed and put back in order to clear the honey from the other side. Though the automatic reversible machine costs more and is heavier, it is far more satisfactory on the whole, if there is much honey to be extracted.

EXTRACTING-FRAMES

The frames used for extracting honey are in form similar to those which hold the brood, except they may not be so deep. However, most bee-keepers use both supers containing the shallow extracting-frames, and also those filled with frames of the full depth. The bees will go into the shallow frames more readily than into the deeper ones, as they are better able to keep the small chambers warm. But if the colony is very strong and the harvest good, the deeper frames are acceptable to the bees and save the time of the bee-keeper. (Plate XVIII.)

WHEN TO EXTRACT HONEY

Some producers practice extracting the honey before it is capped, so as to save the trouble and expense of uncapping. There is one danger attending this method: the green, unripened honey is thus often extracted, and it is the most insipid of sweets. Honey needs to ripen slowly in a warm temperature in order to be palatable. Some, like Quinby, advocate the ripening of honey in vats or evaporators after it has been extracted. But it is the consensus of opinion that honey to be of perfect flavour needs to ripen in the warm, bee-odour-laden atmosphere of the hive. The bees ordinarily leave the honey uncapped for some time as it thus ripens more readily. Therefore, those who produce an especially fine quality of extracted honey usually begin to tier up as soon as the super is fairly filled and before the honey is capped. The bees have ample room to go on storing honey in the interpolated super, and do not bother to cap the honey already stored above. Thus these supers, three or four or as many as practicable, are left on the hive until the end of the honey harvest, and thus the honey attains its proper ripeness and flavour.

There are others who claim that honey is never properly ripened until capped, and therefore practice tiering after the cells of comb are at least partly covered.

UNCAPPING. (Plate XIX.)

There are various knives invented for this process, the Bingham uncapping-knife being the favourite. It is used thus: The frame containing the honey standing on one end and leaning over a receptacle for the caps is held with the left hand, the knife in the right hand begins at the bottom of the comb and running backward and forward as it is moved upward shears off very neatly the covering of the cells. The knife must be very sharp, and skill in cutting is shown in just the merest film of wax which is removed. A pan of hot water should be at hand on an oil stove perhaps; every time a sheet of capping is removed, the knife needs to be scraped on a stick, which will not dull it; and quite often it should be dipped in the hot water to clean it. If there is much uncapping to be done, it is best to have two knives, keeping one in the pan; for cleanliness and heat are quite as potent factors as is sharpness in making the uncapping knife effective.

There are on the market uncapping-cans, the Dadant being the most popular. It is a double can with an arrangement on top convenient for holding the end of the frame on a pivot and with wooden cross-pieces on which the knife may be wiped. Below there is a wire screen for holding the cappings, with a space in the bottom of the can for the honey which drains off, and which is always of the most excellent and delicious quality. Of course, the cappings are saved to be made into beeswax.

CARE OF EXTRACTED HONEY

Honey, whether in the comb or out, will crystallise when subjected to low temperature or when left standing for a long time. However, extracted honey crystallises much more readily than that which is in combs; and this crystallisation is one of the problems of putting up and marketing extracted honey. To prevent it extracted honey should be evaporated until it is thick, sealed while hot in airtight cans, and kept in a room the temperature of which never falls below 65° F. It should be kept in tin or galvanised iron cans, rather than in wood. Some people seem to have been successful in using wooden vessels for holding the honey after having given them a coating of wax; but the way honey gets through small places is more proverbial among bee-keepers than is the ability of the stingy man to do the same; even when a tub or pan seems water-tight the honey will triumphantly work its passage through.

In order to preserve extracted honey in packages it must be canned or bottled, and the air entirely excluded. There are two methods of accomplishing this: First, the honey is heated in bulk, and run off into hot cans or bottles. Second, the honey is put in the bottles first and then heated; in both cases the honey is heated by hot water. Perhaps the easier method is to heat it in bulk, and if there is not at hand a double boiler, one can be improvised by using a wash-boiler in which pails containing the honey may be set. In any case the honey and the water surrounding it should be of the temperature of the room to begin with; then a slow, steady fire is needed to bring the temperature of the water up to 160° F. Mr. Root advises the use of a gasoline stove for this purpose as the heat may thus be carefully regulated, and it is very important that the process be a slow one. After the temperature of the

Taking off upper story of hive containing combs for extracting. The bees have been removed from the supers by the bee-escape seen at the right.

Photographs by Verne Morton

Extracting room showing extractor, strainer, uncapping can, bee-escape, smoker, uncapping knife, gloves, funnel, supers and sections, extractor pocket containing extracted sections, partly filled sections as yet uncapped, extracting frame with comb, etc.

PLATE XVIII

Uncapping comb before extracting the honey.

Photographs by Verne Morton

Placing uncapped comb in one of the pockets of the extractor.

PLATE XIX

honey reaches 160° F. it may be poured into freshly scalded cans or bottles and sealed, air-tight. If bottles are used, the corks should have paraffine poured over them so as to make sure of excluding the air.

PACKAGES FOR EXTRACTED HONEY. (Plate XXI.)

Mason fruit jars are extensively used for this as they are practical, cheap and useful afterward. The No. 25 honey-jar, somewhat resembling the Mason can, is made purposely for putting up honey, and is attractive in appearance. The Muth bottles are made in four sizes for extracted honey; the largest holding two pounds, and the smallest a quarter of a pound. These bottles are decorated with a design of an old straw skep, and bear the prophetic inscription "pure honey" moulded into the glass. Jelly-glasses are often used and paraffined paper is placed over the honey, just as it is placed over jelly to exclude the air before the tin cover is put on. Glass packages are by all means the most attractive for extracted honey in small quantities. However, tin pails of various sizes are in use, and may be serviceable for a cheap and inferior grade of honey, which is thus made ready for the consumer who is willing to buy "aside unseen." But a fine quality of honey rejoices in the light of day and in the scrutiny of eyes which may look at it first critically and then longingly.

The most practical packages for shipping honey in quantity are the large, square cans in common use which hold sixty pounds each. These are convenient for shipping and for measuring and are safe from breakage.

The advantages of extracted honey over that of comb-honey are: Almost double the honey may be produced, because the bees having no comb to build, devote their energies to storing honey; and also the honey from the brood-combs may be extracted; it is more easily and safely handled and shipped; it is more easily produced, since the bees work more readily in the emptied cells of the extracted combs than in sections where they are obliged to build new cells; swarming is more easily controlled, because the bees more readily accept the enlargement of their quarters when the supers contain fully made comb, also larger hives may be used.

The disadvantages are: It is more "mussy" and requires special apparatus; and unless great care is given, the bees will be starved through this convenient way of pilfering their stores.

MAXIMS FOR THE PRODUCER OF EXTRACTED HONEY

Use glass or tin, rather than wood, for honey receptacles.

Be careful not to expose the honey as you take it out of the supers, lest the bees begin robbing. Honey should be canned while hot, and kept from the air.

Heating honey to a higher temperature than 160° destroys its flavour.

Cooling honey to a temperature below 60° produces granulation.

Work in a warm room.

Have a label of your own, with some unique and individual design, which, when placed on the package, will render it attractive.

CHAPTER XI

POINTS ABOUT BEESWAX

HOW TO MAKE IT

Beeswax is a unique product; the little socialists of the hive have formed a trust for its manufacture, which for many and good reasons has never been infringed upon. The special value of beeswax is that it retains its cohesiveness and ductility, under both higher and lower temperatures, than do other kinds of wax. It has a specific gravity of 960–972, and melts at 143–145° F.

Beeswax has been in demand in the world since ancient times, being put to many and diverse uses. Now, however, the making of foundation-comb is the most important of these uses and most affects the beeswax market of to-day. Countless thousands of sheets of wax-foundation are manufactured yearly in America. When foundation was first used it was thought that it might be adulterated safely with parafiine, but it was soon found that beeswax thus adulterated, when subjected to the heat of the hive in summer, would invariably bulge, buckle and sag. We know an apiarist who lost money, honey, bees and temper through trying to use this kind of foundation, which did not stand firm.

Photographs by Verne Morton

Drawing honey from extractor. The honey is strained through a fine wire strainer set in the top of the pail.

Pouring extracted honey into keg for shipping in bulk.

PLATE XX

Extracted honey in pails—candied

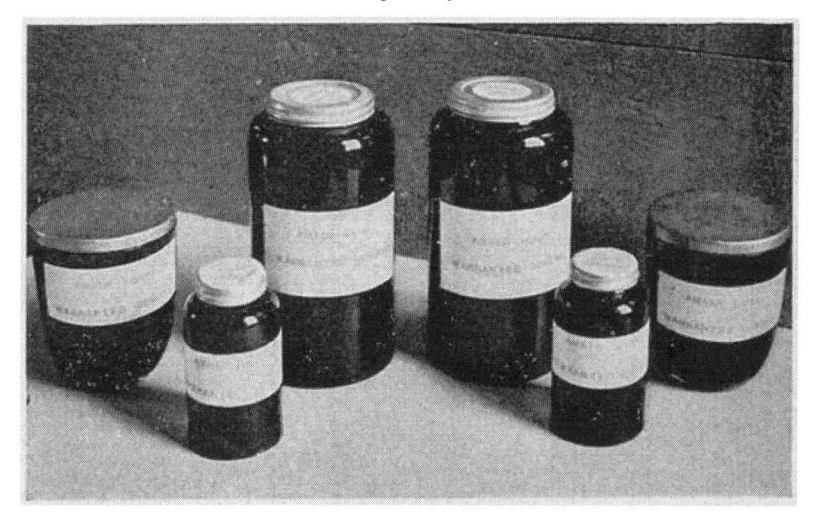

Photographs by Verne Morton

Extracted honey in glass jars ready for market.

PLATE XXI

THE PRIMITIVE METHOD

The process of making wax from honey-comb may be primitive and yet quite successful. Well do we remember the method as practised in the days of our early youth in the kitchen of the old farm-house, usually a comfortable and altogether delightful room. Its yellow-painted floor seemed to have caught the sunshine streaming in through white-curtained windows and held it there for us to tread upon. The chintz-covered settee and Boston rockers, and a few widths of bright rag carpet, made one end a cozy sitting-room. At the opposite end the cupboards, their shelves covered with elaborately scalloped newspapers, and beset with orderly dishes and tinware, bespoke the kitchen. Midway between stood the heavy cherry drop-leaf table which revealed the dining-room; in the midst of this Protean apartment stood the cook-stove, polished so that it would put to shame the rosewood case of a piano, bearing on its top the pink-copper tea-kettle singing gayly after its daily bath of cleansing buttermilk, and holding within the roaring fire which ever seemed the spirit and soul of the place. The neatness that held sway in this kitchen was the perfect sort that conduces to comfort, and not to misery. Only at certain periods was this delectable room given over to discomfort and untidiness; these included "wash-days," the days following the yearly sacrificial rite of butchering and the days when beeswax was made.

We were wont to make beeswax as follows: The broken combs were packed in a muslin bag which was weighted by a sinker and hung in the wash-boiler by tying the bag to the poker, which was placed crosswise the top. By this contrivance the bag was completely surrounded by water, which filled the boiler, and yet did not touch the bottom, and so there was no danger of burning; the cloth acted as a strainer and the bag was pressed occasionally; this act being judiciously performed by means of the tongs, which had previously been cleaned. After the wax had boiled out the boiler was taken off and the whole contents cooled, after which the wax was taken from the top in pieces and remelted in a dish set on the back side of the stove so as not to burn, then poured into oiled bread-tins, and thus caked for the market. Some of it was clarified for special use thus: It was allowed to simmer on the back of the stove for some hours in water to which vinegar had been added, and then was dipped off into scalloped patty-tins and a neat little loop of cord inserted at one side. These cakes when cooled had a long though scarified existence ensconced in work-baskets with spools of thread, for wax thus cleaned and prepared made very pretty little gifts for lady-friends. But the experience of rendering wax was never complete without spilling some of it on the stove, which spread with fearful rapidity and smoked with stifling smoke; because of the certainty of this accident we always made the beeswax in the late fall, lest our bees regarding the pungent smoke as a direct invitation should come visiting in embarrassing numbers.

Sundry old pieces of rag carpet were spread on the floor around the stove to keep the yellow paint intact from the wax, which was so hard to clean off. We had no benzine in those days, and our only resort was boiling hot water, which cleaned off the paint as well as the wax.

One privilege that was always granted to us children on this day was that of having "our fingers made." As the wax was cooling the finger was dipped in it, and the film was cooled while the finger was held very still; then the film was slipped off, a crucial point in the process, and used as a mould into which was poured the cooling wax; and presto! there was the finger as natural as life to every crease and wrinkle, but with a death-like pallor that rendered the row of fingers thus made a fascinatingly gruesome collection, as if they had been chopped off with a hatchet.

This old process of rendering wax in the wash-boiler is still practised where apiaries are small. Mr. Root advises the following modification: Sticks are placed crosswise the bottom of the boiler on which the bag is placed; the bag is packed very full of wax by pressing the comb into balls with the hands before it is put in. Water is added and the whole is placed upon the stove and brought slowly to a boil, then a board that acts as a follower on the bag is placed on top with a heavy weight upon it; this acts as a press and decreases the bulk in the bag, leaving the wax floating on the surface.

THE SOLAR WAX-EXTRACTOR

There are many modern and up-to-date methods advised for extracting wax. The most common is through the use of the Solar wax-extractor, which was invented for extracting honey in California, where the sun can be depended upon to do its work unflinchingly day after day. There was more than myth in the story of Icarus who fastened his wings on with wax, and then dared to face the sun. The ancients eadently knew that no other substance of the sort is so susceptible to the sun's rays. I shall never forget my amazement at the efficiency of the first Solar extractor that I ever saw; it was homemade and there was naught in its appearance to indicate its power. The comb was hard and blackened and full of dirt, while the wax that oozed out and hardened below was as shining and yellow as if the sun itself had exhaled it. A Solar extractor ought to be in every apiary where twenty colonies or more are kept, and into this every fragment of comb should be put instead of storing it to become infested by the bee moth, or leaving it around to incite the bees to robbing. The fragments thus are saved and without any expense or trouble are made into a beautiful product for the market.

There are several of these Solar extractors made and sold by dealers. The Doolittle, the Rauchfuss and the Boardman are the three commonly used. The Doolittle is small and all right for a small apiary. The Rauchfuss has a clever arrangement by which the wax in flowing out overflows from one pan to another and thus cleanses itself automatically. The Boardman is especially adapted for large apiaries, and is on rockers so that it may be tilted to face the sun. The general plan of the Solar wax-extractor is as follows: It consists of a shallow box lined with sheet-iron on which is a frame for holding the comb with a strainer below it, and a place where the wax thus extracted is received. The box has a tight-fitting glass cover, and all the woodwork on the box is painted black so as to absorb all the heat possible. The box is tilted so as to get the direct rays of the sun, and it is important that the cover be of one pane of glass, or several panes matched without cross sash, as such sash interferes with the rays of the sun. Here in the East this extractor works excellently during the summer months. If the wax does not look clean as it comes from the extractor, it may be put through again. The only difficulty with the Solar extractor is that here in the East it works only in the summer time, and that it does not extract all the wax from the refuse which bears the graphic and euphonious name of "slumgum."

THE SWISS WAX-EXTRACTOR

There are many patent wax-extractors which are run by the heat of a stove. The best of these utilise steam for heating the wax, though some of them use hot water in a sort of a modification of the old wash-boiler method. The best of these machines for a small apiary is the Swiss extractor, which may be set over a kettle of hot water, like an ordinary vegetable steamer, which it resembles. The comb is placed in a wire basket, which has a cone-shaped bottom, over which the wax flows down as it melts and drains off through a spout into a pan in which it is to be caked. This machine costs only three or four dollars and is simple and excellent; though it takes but comparatively little comb at a time, it keeps up a continual performance and there is no danger whatever of burning the wax. The basket may be replenished from time to time, and a large amount of wax may be extracted in a day while other work is being performed in the room. Mr. Root has an improved Swiss extractor and so has Mr. D. A. Jones, and both of them are most satisfactory. Mr. Jones's machine is the larger and may be used as an uncapping can as well; the cappings when taken off falling directly into the extractor.

THE WAX-PRESS

Wax is such a precious product in these days of the manufacture of foundation-comb that every particle of it should be saved. This is quite impossible with any of the extractors; as the slumgum always holds much wax, to extract which a wax-press is needed. All wax-presses are necessarily rather expensive machines when bought, and not very easily manufactured at home. In some of them the comb is melted by steam and in some by hot water; in others the heated wax is dipped from a kettle of hot water and poured into the press.

The German steam wax-press is in general use in America, Mr. Root having an improvement on it. It is a strongly built can, at the bottom of which is a place for water, and above it an arrangement to receive the wax drippings. A basket of perforated metal holds the comb and occupies the larger and upper part of the can; a follower worked by a screw presses down upon the heated comb and forces the wax down and out through a spigot. This machine weighs sixty pounds and costs fourteen dollars, and if the apiary contains more than forty hives of bees, such a wax-press will pay for itself soon with the wax which would otherwise be lost.

We have seen one home-made press constructed from a half barrel worked by the machinery of a cheese-press. The follower was cleated on the under side, and the barrel was filled with boiling water; the wax as it was pressed out was run off by a spout at the top. Another which was used successfully was simply a box with cleated bottom and a cleated follower into which the hot wax was poured and pressed out most successfully. One of the wax presses used by many is called the Hatch-Gemmill press, which is run on the principle of dipping the hot wax off hot water and squeezing it through the press.

REFINING WAX WITH SULPHURIC ACID

On a small scale this may be done in an agate or porcelain-lined kettle. Mr, Root even advises on occasion the use of a large iron kettle. The kettle is filled half full of water, 100 parts to one part acid, and is brought nearly to the boiling point over a slow fire; the wax is then added and is kept hot for a little time after it is melted, and then the fire is allowed to die down; as soon as it is cool enough so that the dirt has settled, the wax is dipped off, great care being taken not to stir up the settlings. If an iron kettle is used it should afterward be thoroughly washed with boiling water, and rubbed with fresh lard or some other unsalted grease to stop the action of the acid upon it. Beeswax may be bleached by exposing it in thin sheets to the sunlight.

ADULTERATED BEESWAX

Dishonest dealers have attempted to adulterate beeswax with several substances; tallow, paraffine and cerasin being more commonly used. Tallow or other greasy adulterants may be detected by the smell; and because the cakes of wax containing them feel and look greasy. But paraffine and cerasin are not so easily detected by the eye or feel. The specific gravity test is the one used by dealers. A piece of wax known to be pure is placed in a jar of water, and enough alcohol is added so that the wax will just settle to the bottom. Then a piece of the suspected wax is placed in the jar, and if it contains either paraffine or cerasin it will still float. Another test but not so exact is made by chewing the wax; if it is pure it will be brittle and break as it is chewed, but adulterated wax is cohesive like gum.

MAXIMS FOR BEESWAX MAKING

The wax-market is always good, and the wise bee-keeper saves every scrap of this precious material.

Do not be mussy when making beeswax, or it will take longer to clean up than to make the wax.

Do not use galvanised iron vessels for boiling wax as the quality is thus injured.

Clean wax off with hot water or benzine.

CHAPTER XII

FEEDING BEES

WHEN TO FEED

At least twice during the season bees are likely to need more food than they can get in the fields, if the bee-keeper is to do a profitable business. Once early in the spring when for some reason the nectar supply fails, and it is desirable to stimulate the rearing of brood; again, late in the season when the colony has not enough honey for winter use. When the cupidity of the bee-keeper leads him to extract too much honey, then must he forsooth open his pocketbook and buy expensive sugar to feed back to those whom he has robbed. However, bees should be watched closely; they may need feeding at any time, for it is hard to predict when the honey or pollen harvest may fail in a given locality.

When food is given the bees in the spring, it is largely for the sake of stimulating them to extra activity; and thus develop large, strong colonies ready for work as soon as the harvest occurs. The brood-chambers should be closely watched in the early spring and if there is not sufficient food for the brood present, it should be provided. In the fall the hive should be examined by the middle of September or the first of October. A colony of ordinary size ought to have at least thirty pounds of capped honey. The ordinary Langstroth frame, when filled on both sides, contains about five pounds of honey; therefore, there should be an equivalent of six filled frames in each hive. If the swarm lacks this amount, an estimate should be made of how much more it needs, and this amount should be fed.

Bees are usually fed upon honey or syrup made from the best granulated sugar, although some have claimed that the best grade of coffee-sugars make a good syrup; but the consensus of opinion is in favour of the granulated. The syrup is made in two ways: First, by heat. Melt the granulated sugar in its own weight or measure of water; it should be heated slowly, and never reach a temperature higher than 180° F. lest it burn, for scorched syrup fed in winter is as fatal to bees as so much poison. The mixture should be stirred until the sugar is entirely dissolved, then allowed to cool slowly, and it is ready for use. If there are many to feed, a wash-boiler is a very convenient utensil to use, as it is easier to make a large quantity at a time. Because of the danger of scorching a cold process has been evolved. It consists of taking equal measures of sugar and water; the latter should be boiling hot and the two stirred together until the sugar is dissolved. This may be done in a churn or in the honey extractor. In following this process it is best to add the sugar a bowlful at a time, while stirring the mixture industriously. The syrup should be thin when finished, as it is better to let the bees attend to the ripening of it.

HOW TO FEED

There are two general plans for feeding bees. One is to place the syrup outside the hive, and the other to place it within the hive. The first is much more convenient for the apiarist, but unless the work be done very carefully or in the evening, and the syrup well guarded, the bees may become demoralised and begin robbing. Feeding outside the hive can be done only during warm weather. There are several simple feeders in which the syrup is placed at night, and taken away in the morning; but the method most generally followed is to fill a Mason fruit can with the syrup and place on it a perforated cover, then invert it in a box in front of the hive; the entrance to this box is so connected with the entrance to the hive that robbing is impossible. The box and cover are sold under the name of the Boardman feeder. As there is very little air in the can, the syrup oozes out very slowly through the perforated cover, and the bees take it as fast as it comes. This feeder is satisfactory in that we can tell at a glance when it needs replenishing.

However, most apiarists follow the custom of feeding within the hive, and strive to accomplish this without loss of warmth in the brood-chambers, and without disturbing or daubing the bees. Of all the devices for feeding within the hive, the division-board feeder is the most practical. It consists simply of a division board made to hold the syrup, which is placed in the hive instead of a frame Plate XII. There is a hole at the top so that it may be refilled by simply pushing back the cover, and pouring in the syrup from a pitcher. The only objection to this feeder is its size, as it does not hold more than two pounds of syrup, and if used for fall feeding would need to be filled many times. This feeder is especially useful for stimulating the bees in the spring, and is also most practical in developing nuclei. In a small apiary it is quite practical for all purposes.

Of the larger inside feeders the Smith, the Heddon and the Miller are generally used. These are alike in one respect; they are flat boxes placed directly above the frames and beneath the quilts. The Heddon and the Miller each take a certain specified number of pounds of syrup, so that when we use them we can tell just how much we are feeding. The Miller is especially convenient in this respect, and has one advantage over the others in that the entrance for the bees is directly above the centre of the brood-chamber, so that the bees may enter it easily without loss of heat. This fact renders it an excellent feeder for cold weather. Some still use the pepper-box feeder, which consists of a tin can with perforated cover, inverted above the frame, but this lifts the quilts and lets in the cold, and is awkward to use; and as it does not hold very much it is quite inconvenient to manage.

Some altruistic people take the frames of comb from which honey has been extracted and fill the cells with syrup. This is done by laying the comb flat and letting the syrup into it through a fine sieve, or by using a force-pump with a spray nozzle. After the frame is filled it is allowed to stand on edge until the drain has ceased and then it is hung in the hive, and presto! the bees never know that they have been robbed.

In the happy days, when we were getting our first experience, we fed some colonies for the winter by introducing chunk honey into the bottom of the hive, and it worked like a charm, except that we were obliged to lift the hive to put in the honey, and again to remove the beautifully cleaned comb. One never realises how beautiful empty honey-comb may be unless he has had the privilege of examining a freshly made comb or one which the bees have cleaned. Bee-books advise putting in the chunk honey above the brood-frames, using Hill's device above it so it will not be crushed by the quilt. We have done this, setting the comb in every direction, and our bees ignored it in a most provoking way; but when they found it at the bottom of the hive, they carried it up at once. We never knew why our bees were so contrary in refusing to take the honey from above because other people's bees seem to like it administered in that way.

HOW AND WHEN TO FEED CANDY

If necessary to feed the bees in midwinter many people use candy. This is made by boiling granulated sugar in a double boiler until it is brittle when dropped in cold water. It is then taken off and stirred and poured into flat dishes to harden, from which it can be taken as a cake and placed on top of the frame at the centre of the hive. Some pour the candy into the wooden butter-trays and after it hardens invert the tray over the middle of the brood-nest. Some, instead of caking it thus, mould it in a brood-frame by holding the frame flat on a table or board covered with oil paper and pouring the candy in; it thus hardens fast to the frame, and may be put directly in the brood-nest. The famous Good candy, so called because it was invented by Mr. Good, although the excellence of the product would have given it that name anyway, is made by taking extracted honey, and heating it until it is quite thin, but not allowed to boil, and mixing into it confectioner's sugar until the spoon can no longer stir it; then the mixture is taken out, and placed on a board and more sugar kneaded into it until it is of firm consistency. In hot weather more sugar is needed than in the winter. In making this candy skill is evinced in getting it as soft as possible, and yet stiff enough so that it will not flow. Only the best honey is used for this; and if the confectioner's sugar seems impure, then granulated sugar should be pounded in a mortar or rolled under a rolling-pin until fine and used instead. The confectioner's sugar may be tested by putting a little in a glass of water and noting if there is a sediment.

FEEDING RYE FLOUR

This is given as a substitute for pollen, and is often of great use in the spring when the flowers are late in blossoming, or when severe rains wash the pollen from the fruit bloom. Pollen or its equivalent is absolutely necessary for rearing the brood. The unbolted rye flour, or even oatmeal, or whole-wheat flour may be used by the bees as a substitute with perfect success. The meal may be mixed with the candy if it is desirable; but the usual way is to place it in a trough or box that is shallow, press it down hard so that it will be not more than an inch or two thick, to prevent the bees from getting suffocated while working in it. We must remember that the bee has two rows of holes along each side of the body through which it breathes, and thus could be suffocated as easily in soft flour as in water. The box containing the meal is usually placed a few rods distant from the apiary, and often some old combs with honey in them are placed on top so as to attract the bees to the box, and let them know that it contains food for them. Most bee-keepers say that the box needs to be placed in the sun or the bees ignore it.

FEEDING FOR HONEY

Some bee-keepers practise feeding all the colonies early in the season so that the brood has plenty of sugar-syrup stored near it when the honey season opens; and since the brood-comb is full the bees begin at once to store in the supers. Mr. Boardman who invented the best entrance-feeder paactises this, with the result of getting more honey than other bee-keepers of his neighbourhood, who do not feed at this time. It is especially valuable in years when the honey is scarce, for then the bees store all the honey they gather in the supers. However, there is one thing to consider carefully in this feeding, and that is the relative price of syrup and honey. If the market is glutted and honey is plentiful and cheap, this sort of feeding would better be practised cautiously; but when honey is scarce and dear, it is certainly a safe experiment.

FEEDING BACK

When the sections are not well filled in the late season, it is the practice of some apiarists to feed extracted honey in order to fill them. For this only the best honey is used; it is mixed with water, ten pounds of honey to one of water, and heated so that it is a fluid and then poured in the larger kind of feeders, and is put in at night, as the smell of the heated honey particularly incites bees to robbing. However, the flavour of honey which has been fed back is inferior to that which is only once made, and but a few apiarists practise feeding back successfully.

WATERING BEES

If there is no fresh water in the immediate vicinity of the hives, special provision should be made to secure it, as it is a highly desirable adjunct to a well-regulated apiary. While there are times during the season when the bees get most of the moisture they need from the nectar, there are other times when they drink water eagerly. This is especially so in the spring when they are gathering much pollen and little water, and the weather is warm. Running water is more desirable, and if the drip from a faucet flow over a board, or on pebbles, it affords a nearly ideal drinking place for the bees, since they can drink freely and are in no danger of drowning. Some bee-keepers invert a Mason jar filled with water, on a board that has a few shallow groves, perhaps one-eighth inch deep; the water flows out slowly owing to atmospheric pressure; if a little salt be added to the water the bees lap it up eagerly.

MAXIMS FOR FEEDING

Keep close watch of the bees during the entire season, so as to know whether they need feeding or not.

Feed only good honey or the best sugar.

Never feed scorched sugar in the winter, as it will surely kill the bees.

Observe the practice of feeding at nightfall to preclude robbing.

Never spill the syrup or honey around the yard lest robbers be led on to black deeds.

Feed small amounts to stimulate a swarm or nucleus. Bees are susceptible to small encouragements.

Be careful never to cool off the brood-chamber when feeding in early spring or late fall.

See to it that the bees have water near by, especially early in the season.

CHAPTER XIII
HOW TO WINTER BEES

IDEAL CONDITIONS

THE wintering of bees in the northern latitudes is usually attended with more or less loss. Although we now think we know the conditions necessary for perfect wintering, it is only now and then that they are attained. There are so many unhappy and unpredictable circumstances and vicissitudes, that one must needs be a true prophet as well as a good bee-keeper to be sure that all his swarms will successfully pass the period of snow and cold.

The problem of wintering hinges as much on protection from dampness as on protection from the cold. We all know that double windows in a room keep the frost off the panes. The reason for this is that the dampness of the room is not allowed to come in contact with the cold outside glass. So it is with the bee-hive; if it is single walled the dampness from the breath of the bees causes the frost to gather on the walls of the hive, which later melts and wets the bees so that they chill easily; the double-walled hive is a guard against this condition.

GETTING READY FOR WINTERING

First of all that watchward of the bee-keeper must be fully realised, "Keep the colonies strong." Most men of experience do not attempt to winter a colony that is not large enough to cover at least four of the Langstroth brood-frames. If a colony is as small as this, the division boards should be used to contract the hive and make it as cosy and comfortable as possible. It is far safer to try to winter a colony that covers six frames than one that covers only four; the more the bees the warmer the hive, the less the loss, and also the less missed are those that die. To secure good swarms it is best to keep up the breeding throughout the summer, which can be done by feeding if the honey is scarce.

Next in importance to a strong colony is good honey and plenty of it sealed in the combs, so that wholesome food may help to sustain the bees during this trying period. With a four-frame colony four frames of sealed stores will be enough. A Langstroth brood-frame should hold about five pounds of honey if it is well filled. If the colony is larger, then more honey must, of course, be given. We never allow any of our colonies to begin winter without at least thirty pounds of sealed honey, and when a colony is very large we have given it thirty-five pounds. This may seem wasteful extravgance on our part, but the honey not used in the winter is of use in the spring. It is necessary that the honey be of good quality; the bee is such a neat housekeeper that she will suffer death rather than let food pass through her alimentary canal when she is dormant, and thus render unsanitary the bee-city, a devotion to municipal sanitation which is hardly found elsewhere in the annals of living beings. If honey of poor quality is fed to the bees and they hold within themselves such food, disaster is likely to ensue. In preparing the hive for winter what is known as the Hill's device, which is a series of curved pieces of wood held in place by a strip of tin, is placed above the frames to support the cushion so as to allow the bees to readily climb over them. We use a super cover in our chaff hives instead of the Hill's device.

There are three ways of wintering bees in common use. First in the chaff or double-walled hives left in the open. Second, in tenement-hives. Third, the hives are carried into cellars. The reprehensible way of leaving bees out of doors in single-walled hives with no protection during the winter is no longer practised by civilised people.

OUTDOOR WINTERING

Many apiarists protect the hive by a box, several inches larger than the hive in every diameter, placed over the hive, the spaces between being packed with chaff or dried leaves. A passage-way out is always preserved so that the bees may fly out during the early warm days, and free themselves from the accumulated waste. This is a cheap way of securing the advantages of a chaff hive. Such boxes are sold by the dealers, and many good words are said for this method of wintering.

THE CHAFF HIVE

The chaff hive is probably the most perfect of all of the devices for out-of-door wintering when convenience and the saving of work as well as success are taken into consideration. The chaff hive is a double-walled hive with wall spaces packed with chaff. It is a certain guard against extremes and sudden changes of temperature, as it remains cool in the hot and warm in the cold weather.

The chaff for packing should be fine. That from timothy hay, oats or wheat is commonly used; sawdust and planer shavings and dried leaves closely packed are also often resorted to. The packing should be below as well as on the sides of the hive. A cushion made of burlap and filled with chaff is put above a Hill's device or super cover, as this is the most convenient method of packing the hive on top. The bees should be thoroughly established and have their stores ready as early as October 1st. It is claimed by the admirers of the chaff hive that they prevent spring dwindling by keeping the bees warm in the early spring; and also that they keep the hive cool enough so that the brood is not developed in the combs until the proper time for it.

The chaff hives are sufficiently warm to preserve bees during ordinary winters, but neither they nor any other out-door device were entirely successful during the long, protracted cold of the winter of 1903–04 when many bee-keepers in the Northern States lost 40 per cent. of their bees. It might be wise when such a winter occurs to give some temporary protection to the hives, like covering them with boughs of evergreen or building a close windbreak. The entrance to a chaff hive should always be contracted in winter to keep out cold and mice.

THE TENEMENT HIVE

This is a box made for holding from two to ten hives, and which we have used with perfect success. Our favourite tenement-hive was arranged for six hives in two stories. The bottom was packed well with dry leaves or chaff, and three hives were set somewhat near each other. Entrances were boxed back, affording a front hall for each hive-entrance. After three hives were thus set in and packed with chaff on every side and between, a shelf was put across and on this were set three more hives which were likewise packed. The entrances to the upper row were on the same side as those to the lower row, and the cover of the box sloped back from the front of the hives and was hinged along its highest edge; thus when we wished to examine the hives we lifted the cover and examined the bees from the back side instead of standing directly in front of the entrance. Our losses were rare and small while using these tenements. The advantage of the tenement over the chaff hive is that it is cheaper, and that several colonies packed together help to keep each other warm.

WINTERING IN CELLARS

The way we always wintered bees in the old days was by placing them in a cellar which was used for vegetables and was ill-ventilated and damp. We well remember that in the spring the cellar windows were covered with arrested prisoners; we do not recollect that we lost many colonies, but if we did not, it was owing to the ways of inscrutable Providence rather than our own understanding of the needs of the bees. Probably most of the bees in the Northern climates are wintered in cellars; and because they are wintered in all sorts of cellars with varying degrees of dampness the mortality among them is likely to be great. A cellar fit for wintering bees should be cemented on the floor and sides, made mouse and rat tight, and should be well drained, well ventilated and so arranged that the temperature may be kept in the neighbourhood of 45° F. In such a cellar the hives lifted off the bottom boards should be placed four or five inches apart on two scantlings laid on the floor. In the next tier the middle of a hive should bridge the opening between the lower hives on which it rests. This arrangement gives plenty of ventilation to the hive from below, and it is very important that the air be introduced below rather than above. The cellar should be kept dark, and if the weather is warm and the bees seem uneasy it should be ventilated at night by opening the windows, which, by the way, should have wire screens to keep out intruders. Some leave the bottom boards out on the summer stands, each board bearing the same number as the hive which rests upon it, and thus in the spring it is easy to find the home of each colony; but if the bees are brought into the cellar without the bottom boards on the hive they are quite likely to fly out more or less. Usually, therefore, they are brought in on the bottom boards, and these are piled in some convenient place until needed in the spring. In this case it is advisable to have a map made of the apiary, and the hives and their places numbered on the map, and thus each hive may be returned to its old stand in the spring.

If bees are wintered in the ordinary house-cellar it is far better to partition off the part used for the bees from that used for vegetables, and much pains should be taken to keep the air good and the cellar well ventilated.

Special bee-cellars are in vogue in some large apiaries. The cellar is sometimes made beneath the bee-house, and sometimes it is a structure by itself. Of all such cellars, the Bingham seems to us the cheapest, and surely quite as practical as the others. It is built like a square cistern, twelve feet square at the bottom, sixteen feet at the top and six feet deep; it is cemented at the bottom and on the sides, and the ceiling is flush with the level of the ground. Over this is built a gable roof, the eaves extending down to a drain on either side so that all the water is carried off. The ventilation is secured through a pipe extending from the cellar ceiling to the top of the roof. The floor over the cellar is tight and covered with sawdust; access to the cellar stairs is gained through a trap door. Such a cellar as this should be built on a dry knoll. Mr. Bingham has wintered successfully ninety colonies in this house; and it will hold nearly twice as many.

The bees may come out of the hives and die in great numbers when they are wintered in a cellar. If any such seem distended and swollen they have probably died of dysentery, and the matter should be looked into immediately. However, many of the bees that die in winter are likely to be the old ones which are not vigorous enough to stand the strain of the cold. The cellar floor should be swept several times during the winter and all the dead bees removed so they shall not pollute the atmosphere. The cellar should always be dark, but the bees can be easily examined with a lamp, or what is better, a bull's-eye lantern.

The carrying of hives into the cellar is an onerous task when the apiary is large. The entrance to the hive should be closed the night before so as to be sure the whole family is moved. The hives should be carried with as little perturbance to the occupants as may be; several methods of carrying the hives easily and quietly have been invented. Mr. Root uses a wire bent like a V with the wooden piece of a common pail bale at the angle. The prongs of the V are bent at right angles into hooks which hook under the bottom board; two men carry the hive, one on each side, each with a carrier just described. Mr. Miller has a simple rope carrier that slips under the cleated ends of the hive. Mr. Boardman has a delightful device, which is a carrier in the shape of a push-cart with two wheels. A board just large enough to set the hive upon with rope handles at either side serves admirably.

WHEN TO PUT BEES IN THE CELLAR

This should be done as soon as steady cold weather comes on. In this northern climate the colonies should be ready in October, for the appointed time for putting them in the cellar is likely to occur between the first and fifteenth of November. If put in too soon and the weather is warm they become uneasy; they should be put in during a dry day so that the hives will not be dampened by rain or fog.

WHEN TO TAKE BEES OUT OF THE CELLAR

This is decided somewhat by the bees themselves; if they awaken and push out and try to escape in great numbers, it is a sign that they had best be put out as soon as it can be safely done. Mr. Root makes the practice of putting his bees out of the cellar in the middle of a warm day in midwinter, so that they may have a cleansing flight, and then puts them back in the cellar that night; which shows that a merciful man is good to his bees. However, some other bee-keepers think that this taking them out in midwinter is fraught with danger.

In a climate like that of New York it is hardly safe to take the hives from the cellar before the last of April or the first of May. The general rule is to wait until there is a prevalent temperature of 70°, and the willow, the alder, and the soft maples are in blossom, so that the bees may gather pollen as soon as they are put out. The glowing banners of the red maple blossoms give signal to most of the bee-keepers in the northern climate that it is time that the bees were on the wing.

SPRING DWINDLING

The cause of this is attributed to various conditions by various bee-keepers. The evidence of it is shown by the listlessness of the swarm, and by the dying of the bees. Whatever the reason, all apiarists agree it is more common during cold, backward springs, and that it is less prevalent when the bees are put out in warm, sunny locations. The only remedies suggested are that the brood-chambers be contracted so that the bees can easily keep the comb warm, and that plenty of good syrup and rye flour and water be given to the bees if they are unable to get food from the flowers. Many apiarists have tried the joining of two colonies when this dwindling appears, hoping thus to get enough bees in a hive to keep it warm, but they all agree that this does not help the matter.

MAXIMS FOR WINTERING

Keep the colonies strong. Be sure that a good-sized swarm has at least thirty pounds of sealed stores.

Pollen should not be left in the comb for winter use.

Photograph by M. V. Slingerland

PLATE XXII. WINTER QUARTERS IN CHAFF HIVES

Photograph by Verne Morton

PLATE XXIII. PLUM BLOSSOMS
The fruit-bloom is a great aid to the bees while rearing their brood.

Be sure that the honey is of good quality, and not made from decayed fruit or honey-dew.

Give the hive ventilation from below.

If wintered out of doors, give the bees a chance to fly.

If wintered in cellars, do not put too many bees in a cellar. If you have space for fifty colonies, do not try to winter more than thirty in it.

In cellars take off the bottom boards and arrange the hives so that the bees will get plenty of ventilation from below.

Have a thermometer in the cellar and keep watch of it. This should not show more than ten degrees of variation. If the temperature rises to 55°, open the windows at night.

Keep the cellar dark and the air sweet.

Sweep the dead bees off the cellar floor several times during the winter.

Contract the brood-chamber in the fall, and again, if necessary, when the hives are set out in the spring, until there is only comb enough so that the bees can cover it well.

See that the bees have plenty of food and good water near by when set out in the spring.

CHAPTER XIV

REARING AND INTRODUCING QUEENS

QUEEN-REARING

In a small apiary there is little need for the special rearing of queens; the natural increase may safely be depended upon to supply all the colonies which lose their queens or which have unprofitable ones. It is always well for even the most casual bee-keeper to take the trouble to re-queen from his strongest and best colonies. However, the time when queen cells are naturally built may not be the most convenient or the most desirable time for giving certain colonies a new mother. This being the case, queens may be developed through the power of suggestion, as follows: Select a frame of brood from the best colony; with a toothpick tear down the partitions between three worker cells which contain eggs or larvæ less than two days old and destroy two of the eggs or larvæ; repeat the operation in several places. Place the frame back in the hive, being very sure that there is sufficient space between it and its neighbouring frame, so that good queen cells may be built. If there is a scarcity of honey, feed the bees. The cleverness of bees is clearly proven by the readiness with which they take a hint, and they almost invariably build queen cells upon the comb thus treated.

This method we have found perfectly satisfactory, but for those who rear queens for sale, other very interesting practices have been invented. The greatest of these was devised by Mr. Doolittle, one of the foremost queen-breeders in America. He makes artificial queen cells by dipping a small, smoothly rounded stick in warm wax repeatedly, thus making a little cup, thin at the edge and thick at the bottom. Rows of these little cups are placed on a bar the thickness of a brood-frame and fastened there with hot wax. In each cup is introduced a bit of royal jelly and a very young larva. The bar is then inserted horizontally into a frame of brood-comb, part of the latter being cut away to give room for the future cells, which project down from the bar. In such a royal nursery, he develops his queens for the market.

INTRODUCING QUEENS

Though royalty in the hive is of quite another feather than in human society, yet there is quite as much ado when it comes to installations in one as in the other. While a bee-colony is absolutely devoted to its own queen, it may seriously object to a queen thrust upon it by some outside power. And thus it happens that the introduction of a new queen into a hive is fraught with danger to her majesty as well as to the pocket-book of the bee-keeper.

BALLING THE QUEEN

"Balling" in the hive is an indignity that may well have suggested to other societies the method of black-balling unwelcome seekers after honours. The bees ball an unwelcome queen by gathering around her in a compact mass, remaining there until the unfortunate usurper is smothered or starved, or both. As if to live up to their mathematical reputation, this ball is quite spherical because each bee is an animated atom of centripetal force scrambling and pushing toward the centre. This method of smothering royalty is regarded as an evidence of the worker's reluctance to sting a fertile queen. But observations are recorded which state that the bees on the outside of the ball seem fiercely trying to sting, and that the individuals nearest the queen ofttimes share her fate because of this venomous attitude of their sisters. Whether this use of the sting by the outsiders is for the purpose of fighting their way toward the centre, or whether they are mad with a desire to kill the queen, is by no means a settled question. However, if they were bent upon stinging her to death, she would scarcely be alive after having been balled for some time; while it is a matter of common experience that by breaking up the ball and driving off the murderers, the queen may be saved. Sometimes the bees will ball a queen for a time, then voluntarily release her and accept her.

There are two ways generally followed for dissolving this lump of excited regicides and saving the queen. One is to drop the ball in a shallow bowl of water. This baptism seems to cool the hot blood and the bees swim off, trying to preserve their own lives. The other is to smoke the ball until it dissolves into individual bees, so anxious to get breath for themselves that they forget to shut off the breath of the obnoxious queen. There is a danger attending the latter method, for unless the smoking be done carefully and without blowing hot air on the bees, they will become infuriated by the heat and surely sting the queen; as they evidently regard her, and rightly, as the cause of their suffering.

WHEN TO INTRODUCE A QUEEN

The colony should have been queenless long enough to realise the danger of the situation, but not long enough to have done much toward building queen cells and developing larval queens; in the latter case they prefer a queen of their own dynasty and object to any other. Thus, if a queen is to be superseded she should be removed and, about two days later, the new one should be introduced.

It requires experience to know certainly that a colony has become queenless, for often, when there is no brood or eggs in the cells there is a virgin queen, which eludes the eye, as she does not appear very differently from the workers; a colony with virgin queens of its own cannot be induced to accept an introduced queen.

Mr. Root tests a colony which he suspects is queenless in the following clever way: He takes a cage containing a laying queen and holds it over the frames so that it touches them and the bees may thus get the scent. If the bees have no queen they express their pleasure at this godsend in a very pretty manner by a joyful fluttering of the wings, which conveys the idea of happiness to even our dull senses. To such a colony, the queen may be given with no formalities.

HOW TO INTRODUCE A QUEEN

The colony should be made good-natured by having plenty of food. If there is scarcity of honey, the bees should be fed for a day or so, great care being taken not to start other colonies to robbing by exposing the syrup. The queen is then introduced in a queen-cage, which should be placed between the brood-combs. This should be done as carefully as possible without disturbing the bees. At the end of forty-eight hours an examination should be made, and if the bees are balling the cage, she should be left twenty-four hours longer. When the bees gather around the cage in normal numbers she may be entrusted to them without fear.

QUEEN-CAGES

First of these are the shipping-cages, and it is a thrilling moment when one takes a package from the mail, labelled "Queen Bee, Deliver Quick." And it is still more exciting when the cover slides around, revealing her gracious majesty with a few attendants, safe beneath the wire screen; for no bee-dealer would be so heartless or foolish as to send a queen on a journey without a few ladies-in-waiting to give her companionship and care.

The cage in which a queen is shipped is always tagged or labelled with directions for introducing the queen, which, if followed implicitly, almost always insures success.

The plan of a queen-cage is a cell made of wire screen with twelve to fifteen meshes to the square inch, large enough to allow the bees to thrust in their antennæ and thus get acquainted with their proposed sovereign, but not large enough to permit a sting to be effectively thrust through. The cell, itself, is large enough so that the prisoner will not suffocate if the cage is balled. At one end of the cage is an opening into which is pressed a cork of candy, over which is tacked a piece of pasteboard, through which is a central line of perforations. At first the bees are wild to get at the queen, and incidentally in their attack they get a taste of the candy through the holes in the pasteboard. This distracts their attention, and they work industriously at biting away the pasteboard to get at the candy. And by the time they have worked their way through the delectable door, their attitude towards the prisoner is naturally sweetened, and usually they accept her at once. The "Good candy" is used for this purpose, for the queen is also sustained on this confection during her incarceration, and unless a moist candy is used, she will suffer for lack of water.

Home-made cages are usually employed in introducing queens from the home apiary. These are of various forms and devices, the Miller being a favourite. His materials are as follows: One block of wood 3 x 1¼ x ⅜ in.; two blocks of wood 1 x 7-16 x ⅜ in.; two pieces of tin 1 in. square; two pieces of fine wire 9 in. long; one piece of wire-cloth 4½ x 3½ in.; four wire nails ½ in. long. (Plate V.)

The illustration shows how the material is used. The space between the two small blocks of wood, held in place by the pieces of tin forms a door for the candy. The large piece of wood serves as a plug at the other end of the cage, which may be removed, and the cage set down over the queen, thus capturing her without handling her. When a queen is placed in a cage she should always be allowed to climb up into it. It is not natural for her to climb down.

THE NUCLEUS METHOD

This method of introducing very valuable queens is said by experts to be absolutely safe. It is accomplished by making a nucleus of two or three frames of brood, which is sealed and some of it just breaking through the cell caps. Not an adult bee is permitted to remain, and there should be as few uncapped larvæ as possible, since such will starve. The queen is placed on these combs and the young bees, as they issue innocent of men's scheming, accept her as their legitimate mother, and a colony is soon built up. A nucleus hive of this sort must be placed in a warm room, unless it is hot weather, as there are no bees to warm the brood. Care must also be taken to put a wire screen over the entrance of the hive for a day or two to prevent the queen from escaping if she becomes uneasy midst such a dreary waste of adolescence.

MAXIMS FOR INTRODUCING QUEENS

Be sure the colony is queenless before attempting to introduce a queen.

Be sure the bees have not progressed far in rearing young queens.

Be careful not to anger or disturb the bees by smoke or hot blast or otherwise when placing the queen in the hive.

If honey is scarce, feed the colony before trying to introduce the queen.

Place the cage containing the queen on the frames near the centre of the brood-chamber, wire cloth below her, so that the cage rests on the bars of the frames.

Do not disturb the colony for forty-eight hours after introducing the queen-cage.

Be careful not to allow the queen to escape by flight when liberating her.

Remember, a queen crawls up instead of down.

After queens are two or three years old they should be replaced by young queens.

CHAPTER XV

ROBBING IN THE APIARY

A CAUSE OF DEMORALISATION

THE moral law seems to be as potent among the bees as among men, and if the law be broken there is a moral penalty attached. Nothing demonstrates this more fully than robbing in the apiaries, for as soon as it begins utter demoralisation ensues. All legitimate work is stopped, and all the energies of the bees are devoted to ill-gotten gain and in fighting with each other, or attacking anything or anybody that comes near them. One of the signs of this demoralisation is that robbing makes the bees very cross. The experienced bee-keeper can detect robbing by the angry humming which prevades the apiary, which is a sound as different from the ordinary contented hum of the working bee as martial music is from a pastoral symphony.

WHY BEES ROB

Bees rob for the very human and natural reason that the stores gathered by the hard work of others are coveted and are more easily stolen than earned by labour. When dealing with bees we must always remember that the interest of the individual centres in its own colony, and that it has neither love nor devotion to any other colony, nor to bees in general; indeed, quite otherwise. Endowed with an instinct almost fiendish, bees seem to understand when another colony is weak or discouraged, and, therefore, offers a legitimate field for plunder. Strong colonies are seldom robbed, as they are able to drive out and kill the thieves; but the weak colony is a constant temptation to ill doing, and should be carefully watched and guarded.

WHEN BEES ROB

This occurs usually when there is little honey to be found in the field. Satan provides mischief for the idle six feet and four wings quite as efficiently as for hands. At the end of the honey-harvest there may be a general temptation throughout the apiary to break open vaults of precious stores belonging to others, and escape with the contents. At the close of the honey season strong colonies usually have plenty of sentinels to guard the entrance and look after suspicious strangers. Never at any time should honey-comb be left open around the apiary, for it always leads to robbing. It seems to suggest to the bees that honey gathered is a much more desirable product than that worked for in the fields.

Sometimes when preserving or pickling is going on in the house the bees start to rob the kitchen; and while they may be deterred by screened windows, yet the smell of the sweets may so excite them to desires for forbidden wealth that they seem to become discontented with the grind of daily toil, and so begin robbing their neighbours.

HOW THE ROBBING IS DONE

The robber, unless she be hardened by success, alights on the threshold of the hive which contains coveted stores with an air that is decidedly apologetic, having the general appearance of a prospecting dog with his tail apprehensively between his legs. She shows her guilty conscience by dodging back if she meets one of the legitimate owners coming out of the hive; she is thus trying the skill and prowess of the sentinels, for if by assuming bravado she can pass the sentinels she is usually safe; then she has nothing to do but to fill with honey and to run the gauntlet of the sentinels again, for perchance some of them may stop her and ask her why she is going out of the hive *filled* with honey, a proceeding altogether reprehensible in the bee world. If she succeeds in carrying back to her own hive the honey thus stolen, she creates great excitement there and soon she leads back a mob to pillage; and the air is full of bees bent on wickedness. Mr. Root describes graphically the successful robber thus: "A bee that has stolen a load is generally very plump and full, and as it comes out has a hurried and guilty look; besides it is almost always wiping its mouth like a man who has just come out of a beer shop. Most of all it finds it a little difficult to take wing as bees ordinarily do because of its weight." As a consequence a robber bee borne down with its load of sweetness is likely to crawl up the hive a little way so as to have the advantage of a high place to "jump from" as it takes wing. The virtuous worker comes out of the hive slim and depleted of her load and flies off leisurely to the field, while the robber comes out stuffed full and furtively climbs up the side of the hive in order to be able to be off.

We usually detect robbing in our apiary by the fighting which we observe about the entrance; when we see a pair of workers rolling over and over with each other in the grass near the hive we know that one of them is a robber, but as we do not know which one we are obliged to apply the test meted to the knights of old, and believe that the one who survives is in the right. However, we take measures at once to defend this hive. Also if we discover the bees to be particularly cross some day, we look about to see what has aroused their ire, and nine times out of ten find that robbing is the cause of their ill temper.

HOW TO STOP ROBBING

Contract the entrance to the hive being robbed. To do this we place blocks in front of it, leaving only enough space so that one or two bees can pass in at a time. The robbers having to enter in single file attract the attention of the suspicious sentinels and are either driven back or killed at once. It is unsafe to close the entrance entirely, unless it is done with a wire screen, for the bees within will smother unless the air is admitted through the entrance. It is always well to keep the entrance of the hive which contains a weak colony or a nucleus somewhat contracted as a measure of insurance. If contracting the entrance does not stop robbing, Mr. Root advises strewing long grass about the entrance which he wets thoroughly. The robbers are too wary to try to rob after they have wet their "feathers" in passing through this grass; for thus handicapped they could hardly escape from their enraged victims; while the robbers already within the hive after having thus to crawl out will hasten home to get their clothes dry. Mr. Doolittle uses a common sheet which he places over the hive that is being robbed, while Mr. Miller places over the victimised hive a bee tent of mosquito netting, which has a hole in the top which permits the robbers to escape.

A bee tent is a most useful adjunct to the apiary; there are many times when it may be used, but it is especially useful in preventing robbing. It is simply a tent made of mosquito netting large enough to be set over the hive and the operator. When opening the hives at midday the tent is used to prevent the robbers from attacking the exposed stores.

Mr. Root advises working at nightfall when bees are cross or given to robbing; but most bee-keepers declare that the bees will crawl all over one at night, and are no more to be gotten rid of than a porous plaster, at which Mr. Root promptly responds, "Do not stand near the lamp."

Enterprising men exchange places of the robbers and victims, which produces a confusion that restores quiet. The robbers come back to the weak colony laden with its own stores and join it and help fight off intruders; while a strong colony of robbers is quite capable of defending itself.

Another clever trick practised in Europe is to place some disagreeable, strong-smelling substance on the bottom board of the hive that is being robbed; wormwood, musk, carbolised paper, are used for this. The odour disconcerts the robber unless she is lost to all sense of bee decency; and if she does steal honey from such a hive and returns home her doom is sealed; for although in the bee world the way of the transgressor is not always hard, the way of the citizen that smells differently from her sisters leads to her murder and sudden death.

If a robber colony has almost completed its nefarious work, some people believe that it is best to let it make a clean job, and thus become satisfied there is no more plunder to be had, else the robbers will hunt other hives for depredation. The bee memory seems to be very good, and if the robbers have not cleaned up all the honey which they remember is there, they hunt for it elsewhere. We could never bring ourselves to a frame of mind to permit this calmly, though it seems like sensible advice. Robbing makes us so indignant that we refuse to allow the spoils to the victor.

Bees are certainly very clever, and they are able to learn. The old and successful robber soon reasons it out that where the bee-keeper with a smoker belches forth annoying fumes, there are to be found open hives ready for robbing; such bees will follow the operator from hive to hive, taking their tithe from the helpless colonies. For such robbers as these a way to appease them is a device for letting them rob where they can do little damage. Unsalable comb partly filled is put in hives or supers piled up. These are ventilated by having a wire screen above, the cover lifted, and the entrance contracted so that only one bee can pass in and out at a time. This keeps the robbers busy and happy and out of the way, and the process is called "slow robbing."

Some apiarists remove the robbed colony to a cellar for a day or two until it can recover its communal courage.

BORROWING

When a colony is queenless, or for some reason has no brood, it often allows the robbers to come and go at will, as if it had found life worthless anyhow, and that there was no use in struggling. It seems possessed of the sort of pessimism which leads to stoic recklessness. This can usually be stopped by giving the plundered colony a queen and brood; as soon as the bees find they have something worth while to live for and fight for they are mightily heartened and offer a brave defence.

WHAT BECOMES OF THE ROBBED COLONY

Some of the bees are adaptable, and when they lose courage in defending their own stores they turn about and help carry these stores to the hive of the robbers, which they join, thus swearing allegiance to a new flag. Others, more loyal, will cling to the old empty comb, cluster there, hopeless and despairing until they die of starvation.

MAXIMS TO PREVENT ROBBING

Be sure that the bees are robbing, before applying remedies.

Keep the colonies strong.

Keep watch of the hives in early spring and late fall when there is no honey coming in.

Leave no honey or loose comb open around the apiary under any circumstances.

If bees are determined to rob when the hives are opened by the operator, it is best to work under a bee tent or after nightfall.

Robbing demoralises the whole apiary. If the bees are cross, look out for robbers.

Be very careful in uncapping not to let the bees get a taste of the chyle food, as that makes them very cross and wild to rob.

Keep Italians.

Do not let the colony become queenless, as a queenless colony is legitimate prey for many depredators.

CHAPTER XVI

THE ENEMIES AND DISEASES OF BEES

THE BEE-MOTH (*Galleria mellonella*)

THIS miserable little pest is classic in its devastations as it is mentioned by the old writers, Aristotle paying to it bitter tribute. It belongs to a family of secretive moths which fold their dull wings closely about the body, and thus look more like bits of sticks than like insects. They are called the snout moths (*Pyralidæ*) because the palpi extend out in front of the head in a highly ornamental and striking manner.

The bee-moth is a most insidious creature, hiding in cracks, and, when it flies, darting about with almost inconceivable swiftness. It is only by such means that it eludes the watchful bees. Professor Kellogg observes that the moth simply works against time when it rushes into a hive by laying its eggs rapidly, dodging about with the utmost rapidity to leave as many progeny as possible before the bees can get hold of her and tear her asunder, a fate which surely awaits her. So it seems that even a parasite may be brave and go to certain death in fulfilling its destiny. Mr. Cook says that the moth will lay eggs while her head and thorax are being dissected, which shows that even after death she is efficient for mischief in the hive.

The eggs are small and white and are put into crevices. From such an egg there hatches a caterpillar which spins about itself a silken tube, wherein it lives and in some mysterious way is protected from the bees. It may be that these tubes are of such texture that the bees cannot sting through them; or they may simply be sufficiently thick to protect their inmate from bee observation. The caterpillar lives upon the wax and young bees, and also upon the bee bread; it is a voracious eater, and tunnels through the comb, destroying everything in its path. Those who have had experience with it say that by holding an infested comb to the ear, the noise made by the industrious jaws of the caterpillar can be distinctly heard. Its presence can be detected by the filth and the débris on the bottom board of the hive and also by the silken tubes on the comb. When the caterpillars destroy the bee larvæ, the bees take out the remains and dump them in front of the hive, thus gaining among the ignorant, a reputation for infanticide which they little deserve.

In favourable locations the growth of these moths from egg to adult may require six weeks; the caterpillar when about an inch in length changes to a pupa, in a very thick, protecting cocoon of tough silk. The silk made by these caterpillars is of a most excellent quality; there is in the Cornell University Museum a filmy but strong silken handkerchief made by bee-moths passing and repassing over a flat board; it was made quite involuntarily as the caterpillars spin wherever they go. The bee-moth is especially destructive to stored comb if it is piled close together.

REMEDY

To prevent the injuries of this pest the colonies should be kept strong. The bee-moth follows the rule of other parasites and attacks only the weak and the irresolute, and never injures a comb that is covered with bees. A queenless colony, dejected and discouraged, is usually victimised by it. The Italian bees have learned to cope with the bee-moth and exterminate it whenever attacked by it. Some bee-keepers, when a comb becomes infested, introduce it into the centre of an Italian colony, being sure that the little wretches will find there the fate they deserve. But to us this seems rather an imposition upon a self-respecting colony of bees.

The use of plain, simple, well-made hives is a protection from the bee-moth, as such hives do not afford hiding-places for moths and eggs.

Before comb is stored it should be put in a closed box out of doors, and a saucer of carbon bisulphide placed on top of the comb and left for a day. The deadly gas of this poison is heavier than the air and so falls instead of rising. Care should be taken not to breathe the fumes more than is necessary and hence the work should be done in the open air. Another reason for this is that the gas is inflammable, and hence no fire should be allowed near it. There is another reason why the work should be done in the open and that is because of the sickening stench of the gas. Comb thus treated should be stored in a perfectly tight receptacle, or else be set an inch or so apart on shelves. The bee-moth caterpillar does not seem to like to work in combs that are not set as closely together, as they are in the hives.

If combs infested with the bee-moth are subjected to a temperature of 10° F. the moth is usually exterminated. However, the pest normally passes the winter in the pupa state, and seems to be able to survive in hives left out of doors, wherever the bees can survive. It should be remembered that the bee-moth works only during the summer from May until October, and remains quiescent during the winter. As a matter of fact the modern up-to-date bee-keeper has almost no trouble with the bee-moth. It is a special enemy of the heedless and careless man who neglects his hives, and thus may well deserve to have his bees exterminated.

If a colony is attacked by the bee-moth, the hive should be thoroughly cleaned; new good comb should be introduced and only enough so that the bees can cover it. The infested comb should be fumigated with carbon bisulphide, and after the moths are killed it may be given to a strong colony to clean out and use.

THE INDIAN-MEAL MOTH (*Plodia interpunctella*)

The Indian-meal moth sometimes forsakes its bins of grain and meal and devastates the honey-comb. It does not measure more than one-eighth of an inch across its wings, and its caterpillar is almost too small to be noticed, unless it occurs in great numbers. It sometimes attacks the honey-comb in the North, and in the South it is often a great nuisance. The only remedy for this very small pest is to change the bees to smaller hives, and expose the infested comb to the fumes of carbon bisulphide.

BEE-MOTH MAXIMS

The shiftless bee-keeper is the one who complains of the wax-worm.

Keep Italian bees.

Keep the colonies strong.

Do not leave more comb in the hive than the bees are able to cover.

Use well-made hives with no crevices.

If you see a web upon the comb, hunt out the caterpillar and kill it at once.

If bee-moths get into the honey store-room, close the room and fumigate it with brimstone or carbon bisulphide.

FOUL BROOD

This is an infectious bacterial disease, and its presence in the apiary may be attended by serious results. When it first appeared in America, large apiaries were completely destroyed. In 1874, Professor Cohn discovered the organism which causes the disease, and which bears the name of *Bacillus alvei.* This microbe attacks the immature larvæ and they die in the cells. It always attacks, to a lesser extent, the adult bee; but these leave the hive to die and are not such a dangerous source of infection as are the decaying larvæ. Infected honey is the medium by which the disease is ordinarily spread from hive to hive. Undoubtedly robbing is to a great extent responsible for its prevalence.

HOW TO DETECT FOUL BROOD

The brood appears in irregular patches and it does not all hatch; the caps to the brood cells may be sunken and broken at the centre, the holes being irregular, instead of the neat circular perforations which may be found above the healthy larvae. The sure test of the presence of the disease is found in the dead body of the larva, which is dark and discoloured; and if a toothpick or pin be thrust into it and then drawn back, the body contents will adhere to it in a stringy mass, to the extent of a half or even an entire inch, as if it were mucous or glue; later the bodies of the larvae dry and appear as black scales in the cell bottoms. Another evidence of the presence of the disease is a peculiarly disagreeable odour which permeates the hive, which Mr. Root likens to that of a glue pot.

Remedies.—These have been worked out by many bee-keepers, notably by Mr. William McEvoy, inspector of bees in Ontario, Canada. His remedy is as follows: When there is a good honey flow so the bees will not suffer, all the comb is taken from the infected colony and frames with foundation-starters are given instead. Having no young to feed, the bees use all the infected honey in their stomachs for making comb. At the end of the fourth day all this comb is removed and new-frames with foundation are substituted, and the deed is done.

Mr. Root's practice is to remove the hive from its stand about dark, to prevent robbing, and put another just like it in its place which contains frames filled with foundation. The bees are shaken from the infected hive into the new one. Here they are shut in without food for three or four days, thus being compelled to use all the honey in their honey sacs. Then they are fed and the disease does not appear again in colonies thus treated. Mr. Root burns all the infected comb and frames and disinfects the hives with hot water before they are used again; other apiarists of experience support Mr. Root in this matter. Fumigating the hives with burning brimstone would perhaps be an easier or a surer way of disinfecting the hives.

Many apiarists who do a large business do not destroy the infected comb, but render it in a steam wax press. They also thin the infected honey and boil it for two hours, adding to it a little salicylic acid, and use it to feed back to the bees. But this would hardly pay, unless great care were taken, as one drop of the infected honey would start the disease anew.

Some have tried medicated syrup as a remedy. It is true that syrup made with salicylic acid or betanaphthol will retard the disease, but most bee-keepers believe that it is not a sure remedy.

BLACK BROOD

This appeared in New York so frequently that it was called the New York bee-disease. It was differentiated from foul brood by Dr. Wm. R. Howard, of Texas, who found the bacillus and described it. The chief way of telling it from foul brood is that the contents of the body of the dead larva is jelly-like, instead of gluey. However, Dr. Veranus Moore and Dr. G. Franklin White, of the New York Veterinary College at Cornell University, worked upon this disease for some time and, in 1903, reported that the bacillus of this disease is *alvei* and identical with that discovered and disscribed by Cohn as the cause of foul brood.

PICKLED BROOD

This disease of the brood differs from the others in that the body-contents of the dead larva are watery and that no peculiar stench in the hive emanates from them. Neither is it so contagious as foul brood, though it may seriously cripple an apiary, if not checked. The remedy for foul brood is applied with success to colonies suffering from this disease.

DIARRHOEA

This disease is part of the difficulty of wintering. It is induced by the bees' habit of retaining waste matter in the body until they can fly out of the hive, for thus they preserve the cleanliness of their home. The first symptom of the disease is the soiling of the hive entrance with brownish excrement; the bees are also likely to die in great numbers, their bodies being much swollen.

The cause of the disease is attributed to cold and dampness, and poor food. If, in the fall, the bees store honey made from the juices of rotting fruit or cider refuse, or from honey-dew excreted by plant lice, they are very likely to perish by feeding upon it during the winter. But even with good honey bees wintered in cold, damp hives are liable to contract the disease.

PREVENTION

Give the bees plenty of good food for winter. If the honey they have gathered in August is extracted, feed them syrup from the best of sugar. In winter keep the hives in proper temperature, with sufficient good air and ventilation. After the disease once appears, there is no remedy, except warm weather, which will promptly bring relief.

CHAPTER XVII

THE ANATOMY OF THE HONEY-BEE

A DETAILED discussion of the anatomy of the honey-bee does not fall within the scope of this book; for such a discussion, special works on insect anatomy must be consulted. But there are certain of the more general features of the structure of the bee which the bee-keeper should know; and a discussion of these, merits a place even in an elementary book on bee-keeping.

In treating of insect anatomy it is customary to divide the subject into two parts: first, external anatomy, which treats of the structure of the body-wall; and, second, internal anatomy, which treats of the parts found inside the body-wall.

I. EXTERNAL ANATOMY

The body-wall.—Insects differ fundamentally from man and other backboned animals. With us, the muscles and other soft parts are supported by an internal skeleton; with the insects the body-wall, that part which corresponds to our skin, is hard and serves as a skeleton. In some respects this is a better arrangement than that which obtains with us, for the skeleton of an insect serves as an armour to protect the body as well as a support for the soft parts.

This arrangement of parts holds with the appendages of the body of an insect as well as with the body itself; the legs, mouth-parts and antennæ are all tubular organs, having a firm outer skeleton supporting the inner parts.

Movement of the body and its appendages is provided for by narrow, flexible, zone-like areas in the skeleton which encircle the body and the appendages, at frequent intervals. This segmented condition of the body is easily seen in the hind part or abdomen, which appears to consist of a series of rings.

The microscopic structure of the body-wall is comparatively simple. There is an inner cellular part which consists of a single layer of cells: this is the hypodermis (Plate XXV, 2, *h*); and the outer or hard part: this is the cuticle (Plate XXV, 2, *c*).

The hypodermis is the active living part; it produces the cuticle, which receives additions from it constantly during the life of the insect. On this account, when a section of the cuticle is examined with a microscope it presents a layered appearance.

Moulting of the cuticle.—From time to time during the growth of the insect the outer layers of the cuticle are shed; this is known as moulting. After a moult, the inner layers of the cuticle, which have now become the outer layers, but which are still soft, stretch to accommodate the increased size of the body, and then soon become hard. This moulting, or shedding of the skin, takes place about six times during the development of the bee. Several moults occur during the larval life: one when the larva changes to a pupa, and the last one when the pupa changes to the adult or winged form, just before leaving the cell in which it has been developed.

The head.—The segments of which the body of an insect is composed are grouped into three regions: the head, the thorax, and the abdomen.

The head is the first of the three regions. It is formed of several segments grown together so as to from a compact box. It bears the eyes, the antennæ, and the mouth-parts.

The eyes are of two kinds, which are distinguished as the compound eyes and the simple eyes.

The compound eyes are two in number, one on each side of the head; they are the organs commonly recognised as the eyes. They are called compound eyes because each consists of a great number of little eyes closely pressed together. If a compound eye be examined with a microscope, it will be seen to present the appearance of a honey-comb, being composed of a great number of six-sided elements; each of these is a separate eye.

In addition to the compound eyes, the bee has three simple eyes, or *ocelli*, as they are termed. They are situated on the upper part of the head between the compound eyes.

The antennæ are two slender, many-jointed organs projecting from the front part of the head. Their use has not been fully elucidated. They are doubtless sense organs; and it is believed that certain microscopic pits, which occur in great numbers in their cuticle, are the organs of smell. It is possible, also, that the antennæ function as organs of touch, certain hairs with which they are furnished being the tactile organs.

The mouth-parts are very complicated. They consist of an upper lip, a lower lip, and two pairs of jaws between the lips.

The upper lip is known at the *labrum*. It is a flap-like projection situated above, or in front of, the other mouth-parts (Plate XXV, *u*).

The first pair of jaws, those situated nearest the labrum, are the mandibles (Plate XXV, 3, *md*). Each mandible consists of a single hard piece. They are the biting organs. Certain wild bees, distantly related to the honey-bee, dig holes in wood with their mandibles for nests for their brood. The honey-bee uses its mandibles as tools for the manipulation of wax and propolis, and as weapons in its combats.

The second pair of jaws, which are situated between the mandibles and the lower lip, are the maxillæ (Plate XXV, 3, *mx*). Each maxilla is a long blade. The maxillæ, combined with the lower lip, constitute what, in popular language, is known as the tongue, the organ by means of which the food is conveyed to the mouth, or the nectar extracted from a flower.

The lower lip, or labium (Plate XXV, 3, *l*), is the long central part of the so-called tongue; it bears on

Photograph by Ralph W. Curtis

PLATE XXIV. SUMAC IN BLOSSOM AND BLOSSOM OF MOUNTAIN MAPLE
The Sumac is an excellent honey-producer.

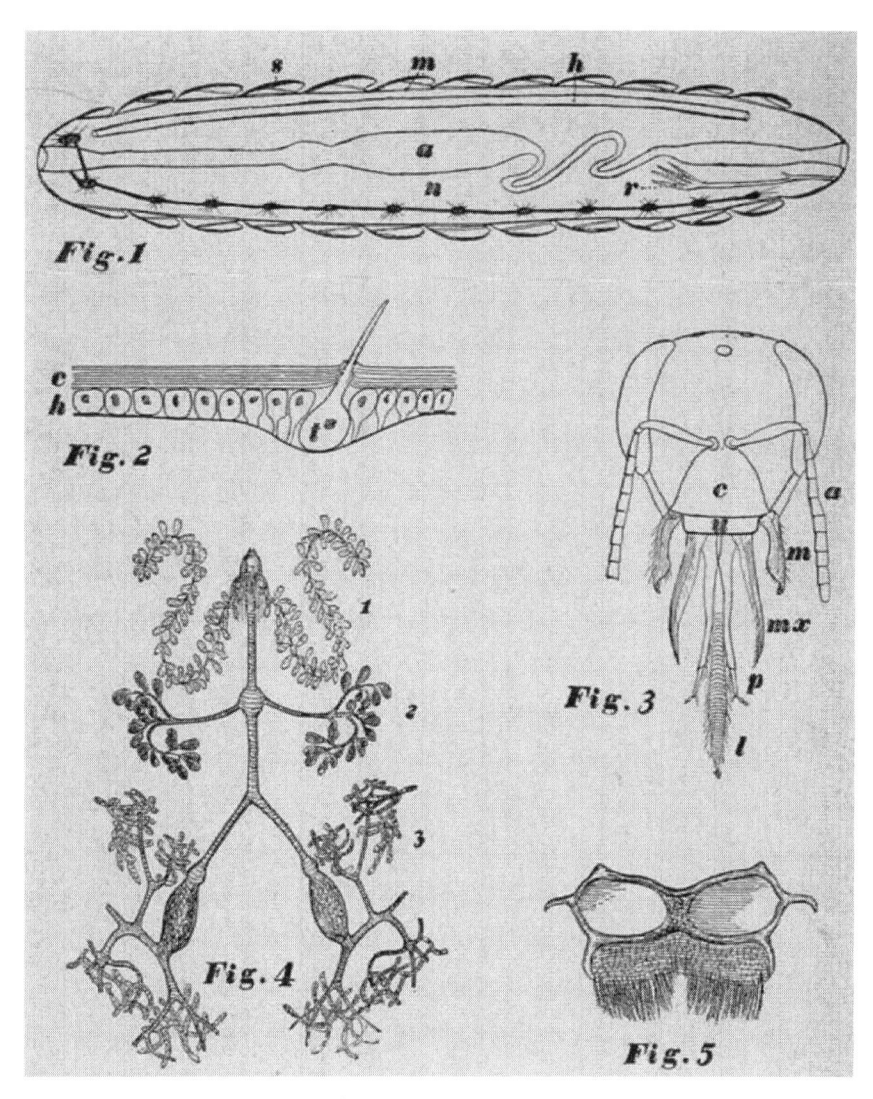

PLATE XXV. FIG. 1.—Vertical longitudinal section of the body of a larva of an insect; s, body-wall or skeleton; m, muscles; a, alimentary canal; h, heart; n, nervous system; r, reproductive organs. FIG. 2.—Section of the body-wall; c, cuticle; h, hypodermis; t, trichogen or hair-forming cell. FIG. 3—Head of a bee and its appendages; a. antenna; c, clypens; u, upper lip or labrum; m, mandible; mx. maxilla; l, lower lip or labium; p, labial palpus. FIG. 4.—Glands

of a honey-bee (after Girard); *1*, supracerebral glands; *2*, post-cerebral glands; *3* thoracic glands.
FIG. 5.—The wax-plates (after Cheshire).

each side a long appendage; these are the labial *palpi* (Plate XXV, 3, *p*).

The thorax.—The thorax is the central region of the body. It consists of three body-segments, which are grown together so compactly in the adult insect that it is difficult to distinguish them. The thorax bears the organs of locomotion, the wings and the legs.

There are two pairs of wings; but the two wings of each side are so closely united that they appear as one. The union is accomplished by a row of hooks on the front edge of the hind wing, which fasten into a fold in the hind edge of the fore wing. The wings are strengthened by a framework of heavy lines, which extend in various but definite directions. Between these lines the wing is a thin membrane.

There are three pairs of legs, a pair borne by each of the three body-segments of which the thorax is composed.

Each leg consists of nine segments and a pair of claws at the tip of the last segment. The first two segments, the *coxa* and the *trochanter,* are short; then follow the two principal segments, the *femur,* or thigh, and the *tibia,* or shank; the five remaining segments constitute the *tarsus* or foot. A striking peculiarity in the *tarsi* of bees is that the first segment differs greatly in form from the other segments and is much larger, approaching the *tibia* in size. This enlarged tarsal segment has received the special name of *metatarsus*.

The legs serve several functions besides that of locomotion. Thus, on each fore leg there is an organ for cleaning the antennæ. The antenna cleaner consists of a circular notch near the base of the *metatarsus*, which is furnished with teeth like a comb (Plate VII, F, *a*), and a spur projecting back from the *tibia* in such a way as to close this notch when the leg is bent. The antenna to be cleaned is drawn through this notch and thus the dirt is combed from it.

On the middle legs there is a strong spur at the distal end of the *tibia* which is used in loosing the pellets of pollen brought to the hive on the hind legs.

The third pair of legs are furnished with three organs which deserve mention here. *First:* the wax pincers. Both the *tibia* and the *metatarsus* are wide; the joint uniting them is at one edge, hence by alternately bending and straightening the leg at this joint, the space between the two segments (Plate VII, B, *wp*) is opened and shut like pincers. This organ is used to loosen from the abdomen the scales of wax. *Second*: the pollen-combs. These are several comb-like series of spines, borne on the inner surface of the *metatarsus* (Plate VII, B, *pc*). When a bee visits a flower the pollen is gathered by the tongue and fore legs and some of it becomes entangled among the hairs on the thorax. It is then combed from these parts by means of the pollen-combs and transferred to the pollen-baskets. *Third:* the pollen-basket. There is a pollen-basket on the outer surface of the *tibia* of each hind leg. It consists of a fringe of hairs, surrounding a smooth, concave area which occupies the greater part of the outer face of this segment of the leg. In it the pollen is packed when combed from the hairs, and transported to the hive.

The abdomen.—The abdomen is the last of the three regions of the body. It consists of a series of comparatively simple, overlapping segments, without conspicuous appendages.

II. INTERNAL ANATOMY

Relative position of the internal organs,—As has been shown in the preceding pages, the body-wall serves as a skeleton, being hard and giving support to the other organs of the body, which are contained within it.

The accompanying diagram (Plate XXV, 1), which represents a vertical longitudinal section of the body of the larva of an insect, will enable the reader to gain an idea of the relative positions of some of the more important organs. The parts shown in the diagram are the following: The body-wall or skeleton (*s*); this is made up of a series of overlapping segments; that part of it which is between the segments is thinner, and is not hardened, this remaining flexible and allowing for the movements of the body. Just within the body-wall, and attached to it, are represented a few of the muscles (*m*); it will be seen that these muscles are so arranged that the contraction of those on the lower side of the body would bend it down, while the contraction of those on the opposite side would act in the opposite direction. The alimentary canal (*a*) occupies the centre of the body and extends from one end to the other. The heart (*h*) is a tube open at both ends, and lying between the alimentary canal and the muscles of the back.

The central part of the nervous system (n) is a series of small masses of nervous matter, connected by two longitudinal chords; one of these masses, the brain, lies above the alimentary canal; the others are situated, one in each segment, between the alimentary canal and the ventral wall of the body; the two chords connecting these masses, or ganglia, pass one on each side of the oesophagus to the brain. The reproductive organs (r) lie in the cavity of the abdomen and open near the hind end of the body. The respiratory organs are omitted from this diagram for the sake of simplicity.

The respiratory system.—The most striking peculiarity in the structure of insects is the form of their organs for breathing, for they do not breathe through the mouth as we do. If an insect be carefully examined, there can be found along the sides of the body, a series of openings; these are the openings through which the air passes into the respiratory system, and are termed spiracles. The spiracles of the honey-bee are small, and are not easily found by one not trained to look for such things; but if the reader will examine the sides of one of our larger caterpillars, he will have no trouble in seeing them. Typically, there is a pair of spiracles, one on each side of the body, in each of the body-segments, but they are lacking in the head and in some of the other segments. The spiracles lead into a system of air tubes, termed tracheae, which carry the air to all parts of the body. When the body of an insect is opened, the tracheae appear as silvery threads, on account of the contained air. In the adult honeybee certain of the tracheae are greatly expanded so as to form large air sacs. (Plate XXVI 6, 2).

The glands,—Those glands found in the body of the honey bee that are of most interest to the practical bee-keeper are the following:

In the larva there is a pair of long, tubular glands, which secrete the silk of which the cocoon is made. These glands open through a common duct, which has its outlet near the mouth.

In the adult worker bee there are four pairs of glands opening into the mouth, which have been much discussed by students of this subject. These glands are designated both by number and by name as follows: System I. or supracerebral glands; system II. or postcerebral glands; system III. or thoracic glands; and system IV. or mandibulary glands.

The supracerebral glands or system I. (Plate XXV, Fig. 4, 1) are situated in the head above the brain. They open by two openings in the floor of the mouth cavity, one on each side.

The postcerebral glands or system II. (Plate XXV, Fig. 4, 2) are situated in the head behind the brain; their outlets unite into a common duct which opens on the middle line of the anterior end of the oesophagus at the base of the tongue.

The thoracic glands or system III. (Plate XXV, Fig. 4, 3) are situated in the thorax; their outlets unite into a common duct, which joins the ducts from the postcerebral glands, the two systems of glands opening through a common opening.

The mandibulary glands or system IV. are two small glands one on each side opening at the base of the mandible.

There has been much discussion regarding the function of these different glands; and even now any statement of conclusions must be regarded as provisional.

The supracerebral glands are large in nurse bees and shrunken in the old bees that no longer nurse the brood; they are normally found only in the workers. It is therefore believed that they secrete the milky food, commonly called royal jelly, which is fed to all larvse during the first days of their development, to the queen larvae throughout their development, and to the adult queen during the egg-laying period. The food fed worker and drone larvae during the latter part of their development is produced in the chyle-stomach of the nurse bees, and is semi-digested food.

The other systems of glands enumerated above produce the saliva, which is supposed to perform a great variety of functions. "It helps the digestion; it changes the chemical condition of the nectar harvested from the flowers; it helps to knead the scales of wax of which the combs are built, and perhaps the propolis with which the hives are varnished. It is used also to dilute the honey when too thick, to moisten the pollen grains, to wash the hairs when daubed with honey, etc."

The wax-glands are found only in the worker. There are four pairs of them. They are situated on the ventral wall of the second, third, fourth and fifth abdominal segments, and on that part of the segment which is overlapped by the preceding segment. Each gland is simply a disc-like area of the hypodermis, the cells of which take nourishment from the blood and transform it into wax. The cuticle covering each gland is smooth and delicate, and is known as a wax-plate. The wax exudes through these plates and accumulates, forming little scales. (Plate VI, X, also Plate XXV, Fig. 5.)

The alimentary canal.—The form of the alimentary canal of the adult honey-bee is shown in Plate XXVI. The following parts can easily be recognised: the oesophagus, a slender tube, beginning at the mouth and extending through the head and thorax to the base of the abdomen. Here there is a sac-like enlargement of the canal, which is termed the honey-stomach; it is in this that the nectar accumulates as it is collected by the bee, and is carried to the hive. Behind the honey-stomach lies the true stomach, the chief digestive organ. Closely applied to the true stomach are several small tubes, which open into it; which are known as the Malpighian tubes; they were named after one of the early anatomists who described them; they are the urinary organs. Next to the true stomach is the small intestine; and behind this, the large intestine.

The reproductive organs.—The internal reproductive organs are situated in the abdomen; there is a set on each side, but the two sets open by a common duct, whose outlet is at the hind end of the body.

The reproductive organs of the female are shown on Plate XXVI (*a*), Fig. 1. There are two ovaries (*o*), one on each side of the body. Each ovary consists of a large number of parallel egg-tubes, within which the eggs are developed. The egg-tubes of each ovary open into an oviduct (*od*). The two oviducts unite and form a single tube on the middle line of the body; this is the *vagina*; the *vagina* leads to the external opening of the system. Communicating with the *vagina*, there is a sac-like pouch, the *spermatheca*, which is the reservoir for the seminal fluid; this is filled at the time of pairing, and the *spermatozoa* may remain alive in it for several years.

Each egg-tube produces many eggs. As the eggs increase in size, they pass towards the oviduct. When the egg is fully developed a shell is formed about it. This shell has a minute opening through it at one end; this is the micropyle. At the time the egg is laid a *spermatozoon* may pass from the *spermatheca*, where thousands of them are stored, into the egg through the micropyle, and thus the egg is fertilised.

With most animals, the egg must be fertilised in order that it may develop. But with bees, both fertilised and unfertilised eggs develop, the former into females, that is, workers or queens, the latter into males, that is, drones.

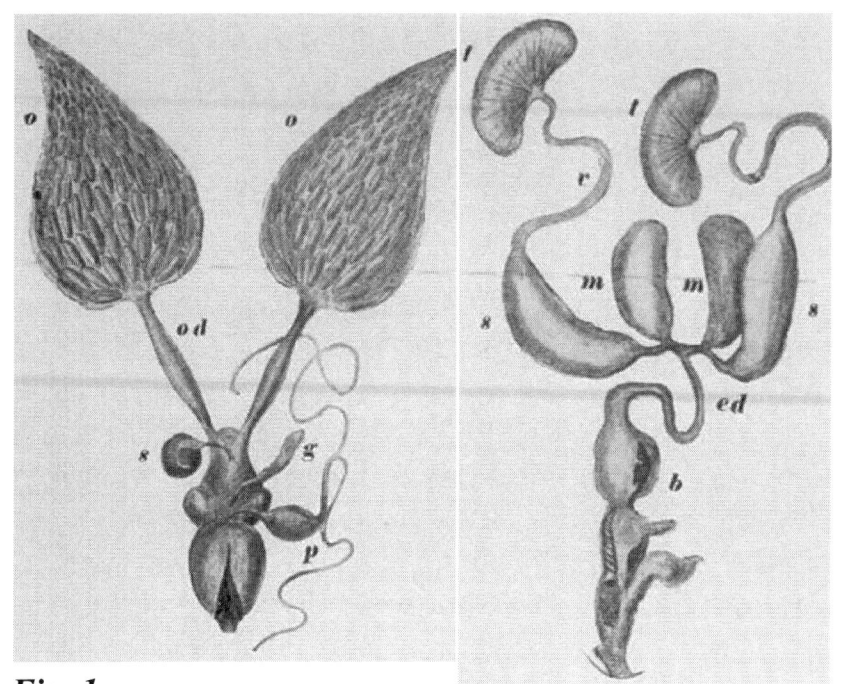

Fig. 1

Fig. 2

PLATE XXVI. (a.) The reproductive organs of the honey-bee. (From Leuckart, slightly modified.) FIG. 1.—Reproductive organs of a queen; *o, o*, ovaries; *od*, oviduct; *s*, spermatheca; *g*, gland; *p*, poison sac connected with the sting. FIG. 2.—Reproductive organs of a drone; *t, t*, testes; *v*, vas deferens; *s, s*, seminal sacs; *m, m*, mucous glands; *ed*, ejaculatory duct; *b*, pouch or bulb. The bulb and the following parts are everted through the outer opening at the time of pairing.

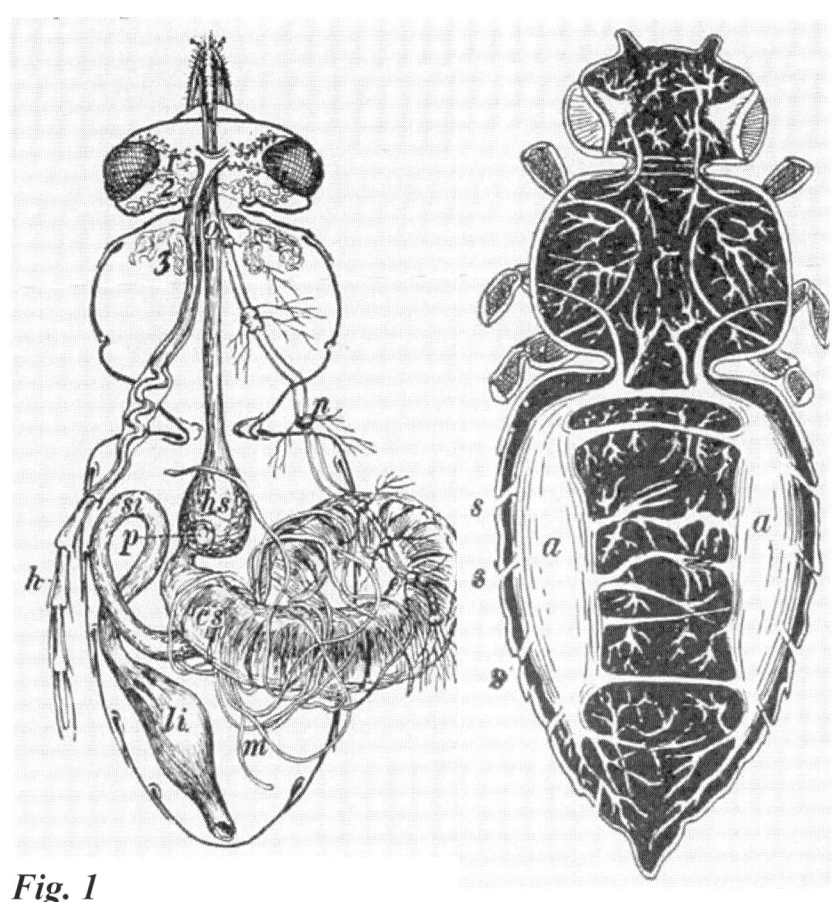

Fig. 1

Fig. 2.

PLATE XXVI. (*b.*) FIG. 1.—The internal anatomy of the honey-bee (after Cheshire), *o*, œsophagus; *hs*, honey sac; *p*, stomach-mouth; *c s*, true stomach or chyle-stomach; *m*, malpighian tubes; *si*, small intestine; *li*, large intestine, *h*, heart or dorsal vessel; *n*, central nervous system; *1*, supracerebral glands; *2*, postcerebral glands; *3*, thoracic glands. The respiratory system and the reproductive organs are not shown in this figure (after Cheshire). FIG. 2.—The respiratory system of the honey-bee. *s, s, s,* spiracles; *a*, enlarged trachea or air sac (adapted from Leuckart).

Courtesy of Ginn & Company

PLATE XXVII. AN OLD-FASHIONED APIARY

Closely associated with the reproductive organs of the female is the sting; this is a barbed dart connected with a poison gland, whose use is well known.

In the abdomen of the male there is a pair of organs, the testes, in which the *spermatozoa* are developed. These correspond in position to the ovaries of the female, but are much smaller. From each testis there extends a tube corresponding to the oviduct, this is the *vas deferens*. The two *vasa deferentia* unite and form the single ejaculator duct. Each *vas deferens* is enlarged just before it joins the ejaculatory duct, forming a reservoir for the accumulation of *spermatozoa*; these reservoirs are termed the seminal sacs. Appended to each seminal sac there is a large glandular sac, hich adds mucus to the seminal fluid. Near the outer end of the ejaculatory duct there is a pouchlike enlargement into which the *spermatozoa* pass. Here they are massed into a compact body, known as the spermatophore, which is transferred to the female at the time of pairing. The terminal part of the reproductive organs of the male, the intermittent organ, has several appendages, which are firmly grasped in the opening of the reproductive organs of the female and are torn from the male when the two pairing individuals separate. This causes the death of the male. The male has no sting. (Plate XXVI, Fig. 2.)

CHAPTER XVIII

INTERRELATION OF BEES AND PLANTS

HONEY-FLOWERS

THE facts revealed by science are not always beautiful, however interesting they may be. But the discovery of the interrelation of flowers and insects, that partnership which has existed so long that it has modified both partners, seems to belong to the realm of art or poetry rather than to that of science. Since plants must needs spend their lives where they are developed from seeds, they may not roam abroad like animals to seek their mates. But this is a difficulty which they readily overcome, through sending their messages by the uneasy, flying insects; and of all these messengers, the bee is surely the flowers' favourite. Its fuzzy body, admirably adapted to be a pollen brush, its swift wings and its sedulous attention to business, all tend to make it the most important of the flower partners. Thus, especially for the bee, have thousands of flower species developed nectar, in pockets placed cunningly to entice her to take upon herself a pollen load. And for countless ages the flowers have painted their petals various hues and shed on the atmosphere their perfume, to advertise to the bee-world that they had pollen and nectar to spare.

This partnership has naturally modified the insects less than the flowers, as the latter were obliged to develop innumerable devices for winning attention from their messengers; naturally also the insects have been more largely modified in their mouth-parts and appliances for carrying pollen, than in other directions. Their habits have also been modified in a measure, and the bee has in some mysterious way been persuaded to work on one kind of blossom at a time. The poetic reproach that the bee is a heartless rover, rifling the lily of sweets only to desert her for the rose, is as unjust as it is untrue. Repeatedly have we watched a bee at work in a bed of pinks. Though clover and other blossoms were near by, she passed methodically from pink to pink, and naught tempted her to fickleness. That the bees use pollen for bread, is a part of the bargain; for the flowers grow it in plenty for both themselves and their partners.

Each species of "honey-plant" has developed its own special device for securing the services of bees to carry its pollen; and no study is more interesting than the unravelling of these flower secrets. Even the novice may do this by asking the flower these three questions: "Where is your nectar?" "Where is your pollen?" "What is the path the bees must follow to get to the nectar?" For ready and accurate answers to such questions, the flowers are not to be surpassed; and if there is any doubtful point, the bees are ready to help elucidate it. There are so many flowers that have become the special of the bees that it is not within the scope of this book to make an adequate list. Some of the more important are the flowers of trees, and some of farm crops; some bloom in gardens and some are mere weeds.

TREES

Fruit trees, when in blossom, give much pollen and honey at a time when these are greatly needed by the bees for rearing brood.

Peach, pear, apricot, plum and especially apple trees, when in bloom are encompassed about with the happy chorus of busy and grateful bees; and no other creatures can so successfully vocalise blissful contentment as they. Many careful experiments have proved beyond doubt that the help of bees is necessary for securing the pollenation requisite to produce good crops of fruit. The wise and successful fruit-grower recognises this fact and, mindful of his true interests, does not spray his fruit trees with poisons while they are in blossom, lest he thereby kill his friends, the bees. Moreover, to use arsenical sprays, at such a time, is injurious to the petals and the fruit-producing organs of the flowers; and it is also too early to spray successfully against the codlin moth. In many states, legislation forbids the spraying of poisons during fruit-tree bloom, because it is a useless and wanton destruction of the bees.

Some time since great injustice was done the bees through the accusation that they punctured the ripe fruit for the sake of the juices. This was the special complaint of grape-growers. Investigations have proved that bees never puncture the rind of ripe fruit, although they sometimes are tempted to sip the oozing juices, after the rind is broken through some other agency.

Even before the fruit-bloom the willows offer a feast to break the fast of the hungry swarms. Half the winter the pussy-willow stands waiting in her furs to be ready with her grist of pollen, so that the bees may make bread during the first warm days of spring. The willows burned their bridges behind them eons ago and depend almost entirely upon the bees for fertilisation, since they are diœcious. Some apiarists have claimed that their bees get no nectar from certain species of willow; but this could hardly be so if trees of both sexes were present in a locality; for the staminate flowers offer pollen and the pistillate flowers give nectar to induce the bees to fetch and carry for them.

The maples are not much behind the willows in offering the bees food after their winter fast. The bloom of the red maple is regarded by most bee-keepers as permission from Spring to bring out the bee-hives from the cellar and tenements. All our common species of maple are very much visited by the bees.

The locusts often yield large crops of honey, although they vary with the seasons in this respect. Honey-locust, when in bloom, is covered with bees.

The tulip tree is one of the most beautiful of our ornamental trees and it gives a great amount of dark, rich honey. In New York it blossoms in May and June and, like the locusts, is a great help to the bees after the fruit-bloom is over. This is a common tree in the woods of the South and is not rare in Northern forests. It should be planted even more generally than it is at present, for the sake of the bees.

The basswood, of all honey-producing trees, is the most important and most beloved by the bees. It blooms in July and only for a brief season; therefore, it is important that the colonies be strong and able to make the most of these few precious days of harvest. Basswood honey has a strong flavour when first gathered. But after it is ripened and sealed it has a delightful flavour. The way our forests have been stripped of basswood during the past twenty-five years is nothing less than heart-rending to the bee-keeper; for to him this tree ranks next to the white clover in importance. Mr. Root had a single colony take forty-three pounds of honey in three days from basswood, and Mr. Doolittle had a colony take sixty pounds in the same period. The tree is beautiful, and might well be used for shade along roadsides and also in ornamental planting. It grows rapidly; young trees, transplanted from the woodland, blossom in five or six years thereafter. No bee-keeper should allow the basswood to be cut on his premises; and he should grow as many young trees as possible.

Other honey-producing trees of note are the

Thorn-apple blossoms.

Photographs by Ralph W. Curtis
Wild crab-apple blossoms.
PLATE XXVIII

PLATE XXIX. BUCKWHEAT IN BLOSSOM

sourwood (*Oxydendrum arboreum*) of the South, the guajilla of Texas, the cabbage palmetto of Florida, and the eucalyptus of California.

The flowers of sumac often yield much nectar and are sedulously worked by the bees. This picturesque shrub is not properly appreciated because it is so common. Its foliage is beautiful in the summer and is brilliant in the fall. Its blossoms, as well as its fruit, conduce to make it an interesting and ornamental shrub for planting.

HONEY PLANTS WHICH YIELD OTHER VALUABLE CROPS

To raise plants solely for the sake of the honey they produce has not proved a financial success so far in America. Mr. Root estimates that it would require 500 acres covered with plants blooming in succession to keep 100 colonies of bees busy; and, at present, most land here is worth too much to be put to such use. It is doubtful if artificial pasturage will ever prove a paying investment in agricultural sections.

However, many apiarists devote some land to honey-gardens, and such a garden may be a beautiful and interesting place, for many of the honey-plants are ornamental. Also, many apiarists have introduced certain honey-weeds on waste land in the vicinity of their apiaries with excellent results.

Fortunately, many plants very valuable to the agriculturist and horticulturist are the best honey-producers; and if a farmer has only twenty colonies of bees, it will pay him to reap one crop from his land and let his bees reap another.

Of all such plants the clovers are the most important. Not only do they make the best of forage and hay, but they also help to fertilise and aerate the soil, and should be a factor in every crop rotation. Clovers and other *legumes* have upon their roots nodules filled with bacteria, which are underground partners of the plant. These bacteria fix the free nitrogen of the air and leave it in the soil available for plant food. Red clover is not so great a source of honey as are the other clovers, since its corolla tubes are so long that usually it is only worked by bumblebees. But the long-tongued Italians are able to get considerable honey from red clover at times.

Crimson clover grew as a weed for a long time in America before it became an important factor in horticulture. It is an annual and its home is in southern Europe. It thrives best in loose, sandy soils and is of great value as a cover crop for orchards. It is a good honey-plant.

Alsike is a perennial and grows in low meadows, from Nova Scotia to Idaho. Its blossoms look like that of the white clover, except that they are larger and are tinged with pink. This is a valuable clover for pasturage, and also for hay, and it stands next to the white clover as a honey-plant.

The white clover is the very best plant for producing honey in the United States east of the Rocky Mountains, and the flavour of its honey is famous the world over. While in hard soil, the white clover lasts only two or three years, it is perennial on rich, moist lands. It is a cosmopolitan plant and may be found in almost all regions of the temperate zones. It is an ideal plant for pastures and should be established everywhere on land not under the plow. It shows well its partnership with the bee by turning down its flowerets as soon as they are fertilised, and leaving those in need of pollen still erect. We have seen a head of white clover with a single floweret, erect and white, calling to the bees, while all of its sister flowerets were deflected and brown.

Among the medics we find the veteran of all clovers, the alfalfa, which has been under cultivation for twenty centuries, and came to America with the Spanish invasion. It was established in California in 1854, and has worked its way eastward. But it is only recently that it has been practicable to grow it in the East. This has been made possible by the discovery that it will grow on soil inoculated with its root bacteria. Alfalfa is a true perennial and may be cut for hay three times a season, and is of highest value as fodder or forage. It is a superb honey-plant, furnishing great quantities of light-coloured and excellent honey. It will support more bees to the acre than any other plant known. For artificial pasturage it is the most promising of all honey-plants.

Buckwheat is, in many localities, doubly profitable as a grain and as a honey-plant; especially is it so in middle and western New York, where the hills in autumn are made brilliant with great fields of the wine-red stubble. Buckwheat is usually sown late in the season, often on ground where oats have already been grown earlier in the year. It blossoms in August and even in September, and furnishes a wealth of nectar when there is little to be found elsewhere. The honey made from buckwheat is dark, reddish brown and brings a lower price in most markets than do the lighter-coloured varieties. Though it is strong in flavour, it is preferred by many, and on our table it alternates with basswood and clover. It has always seemed poetic justice that the plant which produces buckwheat cakes should produce the honey to eat with them. The following are the good points of buckwheat as a crop: It is profitable, the grain always brings a good price; it grows well on poor soil; it is one of the best agencies for ridding a field of weeds. There is a certain gameness about buckwheat which we have always admired and which was thus characterised by a farmer of our acquaintance: "Buckwheat is a gritty plant; if it can get its head above ground it will blossom. I have seen it, during dry seasons, blossom when its stalks were so short that the *bees had to get down on their hands and knees to gather the honey*." While this may be putting the case rather strongly, yet it expresses well the habits of the plant.

Black mustard, rape and turnips all furnish nectar for excellent honey. The seed of mustard and rape brings a good price, and the root of the turnip is always valuable.

The blossoms of the red raspberry yield a delicious honey, and the grower of small fruits may well add bees to his farm as a source of profit.

GARDEN FLOWERS

Most of the blooms in flower gardens, as well as vegetable gardens, are worked by the bees. Mignonette is a valuable honey-plant, as it blossoms for a long time. Marjoram, thyme and sage give rich, spicy honey. The sunflower is also a good honeyplant.

WILD FLOWERS AND WEEDS

We shall never forget our profound amazement when we saw, for the first time, in a narrow valley of the Mojave Desert a great city of white bee-hives. Nothing in that desolate landscape could we discern that bore the slightest resemblance to a honey-plant. The gray sage-brush which grew everywhere looked to us about as promising for honey-production as so much slag from a furnace; and yet this sage-brush of the desert gives the bees the best of pasturage. The bloom begins down in the valley and climbs the mountain side slowly, thus giving bloom for a long period.

There are two species of sage that yield honey, the white and the black, or button sage. They are allied to the mints, which are generally good honeyplants. We learned to care much for the spicy sage-honey. A professor of Greek, who was for some time in the American school at Athens, tells us that the sage-honey is very similar in flavour to the famous honey of Hymettus, which is made from thyme.

The horse-mint is a very important honey-plant of the lower Mississippi Valley and of Texas. Its corolla tubes are so long that only the Italians and other long-tongued bees can get its honey. Catnip, motherwort and gill-over-the-ground and gall-berry all furnish an abundance of nectar.

The blue thistle emigrated to Virginia in colonial times, and now covers with a heavenly blue thousands of acres of the desolate, uncultivated, red Virginia soil. It is a great boon to the bees of the region, as its blossoms creep slowly up its stalks, thus affording nectar for many weeks. It is related to borage, which is another good honey-plant.

Spider flower (*Cleome fungens*), the Rocky Mountain bee-plant (*Cleome integrifolia*), and figwort (*Scrofularia venalis*) have all been planted by bee-keepers in their honey-gardens, because they give such a great amount of honey per plant.

During September and early October the bees work busily on the various species of goldenrod and asters, and gather from them a considerable amount of honey, which is rich in colour and taste. The two common species of *Impatiens* also give the bees good fall pasturage.

Fireweed (*Epilobium angustifolium*) comes wherever forests have been cleared and burned off. It blooms late and yields a fine quality of honey. The unlovely Spanish needle (*Coreopsis*) also gives much honey. The milkweed yields good honey, yet it overdoes the matter by loading the feet of the bee with its pollen sacs, until the poor messenger dies of exhaustion under the burden of its message, or dies a prisoner in the blossom.

The cheerful and ubiquitous dandelion has this much in its favour, that it is beloved by the bees and often gives them honey and pollen at a time in the spring when they need it for brood-rearing.

Of all the weeds which will pay the apiarist to establish in waste places, the most profitable are the mellilots, or sweet clovers, of which there are two species, the white and the yellow. These are most beneficent weeds, for they carry nitrogen to the soil like other clovers, and they are easily exterminated by cultivation, so they are not to be feared.

Sweet clover in blossom fills the atmosphere of the country road with perfume, for it is almost everywhere a roadside weed; and, as might be expected, it is most attractive to the bees. While the honey made from it is rather strong in flavour, yet it is of good colour, and sells well. When it is mixed with the honey from other flowers, it adds much to the excellence of the flavour.

CHAPTER XIX

BEE-KEEPERS AND BEE-KEEPING

Most business occupations lead to rivalry and all the selfish emotions incident to competition; not so is bee-keeping; quite the opposite, indeed, as there is a freemasonry that holds bee-keepers together and renders their attitude toward each other friendly and helpful. Bee-keeping is everywhere a bond of brotherhood and a sign of congenial tastes. One night at a dinner-party the gentleman on my right was a stranger, known to me only by reputation as a lawyer of high standing and great erudition. He was reserved and silent, and evidently bored by the trivialities of table-talk. Some one incidentally spoke of our bees, when the face of my neighbour became illumined with interest, and he said, "I am sure that by becoming a lawyer I spoiled a good bee-keeper. I have never found anything else so interesting as bee-keeping;" and thus was swept away the curtain of cold conventionalism which had hung between us, and we began, from that moment, to be friends.

Nowhere is this brotherly interest more noticeable than in the bee-books and the bee-journals. The former bear evidence, on every page, of kindness and courtesy to all; while the latter are like friendly correspondence published, wherein John Smith of Maine explains his views on the bee business to Timothy Jones of Oregon, and incidentally sends his kindest regards to the family.

I never take into my hands that delightful book, "A B C of Bee Culture," without turning to the biographies of noted bee-keepers, and looking again at the faces there depicted, noting the noble forehead of Huber; the keen, scholarly face of Dzierzon; the judicial countenance of Friend Quinby and the beautiful expression of the venerable Langstroth. And thus on, page by page, and getting, by the way, a friendly greeting from the kindly eyes of Professor Cook, that most excellent of good teachers; and finally deriving sincere satisfaction from a long look at the keen, humorous face of Mr. A. I. Root himself. These leaders in apiculture are men with whom one is proud to be associated. And the fact that there are, in the United States, 300,000 persons engaged in bee-keeping, makes one hopeful that our republican institutions are to be guarded by intelligent citizenship.

It is interesting to note that knowledge of bees has been given to the world by men who have attained the high peaks of scientific fame. Such knowledge began with Aristotle and Pliny in ancient times, and received no additions during the uncertain Dark Ages. It began anew with Swammerdam, in the seventeenth century, was augmented by Linnæus, De Geer, Réaumur, Bonnet, Lyonnet, Fabricius, Latreille, Lamarck, and finally reached a climax in the study of the habits of the bee by the blind Huber, who was born in Geneva in 1750. His observations made with his own brain, but with the eyes of his wife, niece and servant, form a classic in bee-literature. In 1811 there were born, a continent apart, two great bee-keepers: Langstroth in Philadelphia, and Dzierzon in Silesia. Both were clergymen, but were also true scientists, and both invented means by which the combs could be moved and examined. Langstroth carried his invention farther, and in 1852 devised the movable frame which revolutionised bee-keeping.

Up to this time the business of the bee-man was largely guess-work. He did not know anything about the condition of his bees in the hive, for he had no way of penetrating that dark chamber. The ways of reaping the honey-harvest were devious; at best the combs were torn from the hives with little regard for the rights and lives of the bees. Finally, there was devised the truly infernal plan of killing the bees with the fumes of burning brimstone, before taking their treasure; this method undoubtedly originating in the turgid theology of the times.

However, about the time of Langstroth, someone, or perhaps many, had discovered that bees stored their honey in the upper part of the hive; and the old box-hive had a few auger-holes in the top, over which was inverted a box, which the bees usually filled, and thus saved themselves from the brimstone pit. We remember well the delight in our family when we used, for the first time, such boxes with glass sides; and as we saw they were being filled with combs, we rejoiced that we need not "take up" any more swarms, as the suffocation by sulphur fumes was termed.

When, to the invention of the box super, was added the greater invention of Langstroth, and finally thereunto was added the invention of the honey-extractor, bee-keeping became a science, instead of a haphazard avocation.

Bee-keeping in America has since then passed through many phases and survived many experiments. Our bee-keepers have been wide awake and willing to try all things and hold fast to the good. Once, having read of the floating apiaries of the Nile, which follow the flower bloom along the banks, an enthusiast tried the same scheme on the Mississippi River, starting at the southern part early in the season and coming north abreast of the spring; but too many bees were left behind to make this profitable. Another enterprising gentleman took his bees south winters, but the cost of transportation took away the profits. Now, however, another plan for gaining pasturage is proving most successful; *i.e.*, the establishment of out-apiaries. As seventy-five or a hundred colonies will usually take all the nectar of a given locality, the bee-keeper places his surplus colonies in other apiaries far enough distant, so there is good pasturage for all. Mr. A. L. Coggshall has about 3,000 colonies in out-apiaries in central New York.

The up-to-date bee-keeper is not merely an operator in his apiary, but a co-operator with his bees, and we firmly believe that the bees are being educated by the partnership, as well as the beekeeper. Bees certainly do learn by experience, as is well instanced by some Cyprians, which, in their native home, build columns of wax and propolis at the hive-entrance to keep out the large death's-head moth which preys upon them. After living in this country for two years, the bees discovered that there were none of these moths about, and so ceased building these bars. The readiness with which our bees use the comb-foundation and fit their combs to the frames and follow the hint given by the starters in the sections, all point to their adaptability; and however others may regard the matter, we never take off a section filled with just one pound of pearly comb and amber honey that we do not pay tribute to the bee intelligence which placed it there, and regard the little artisans as true partners in our enterprise. And we never doubt that in the future this co-operation and co-education of bees and beekeepers will result in a perfection of honey-production as yet undreamed.

Photograph by Ralph W. Curtis

PLATE XXX. BOX-ELDER
Staminate and pistillate flowers. This and the other maples and the willows give the bees pasturage in early Spring.

Photograph by Ralph W. Curtis
PLATE XXXI. BLOSSOMS OF BLACK LOCUST
The locusts often yield large crops of honey

CHAPTER XX

BEE-HUNTING

THE mere mention of these words always brings to us memories of high hills, wound about by picturesque roads, bordered by rail fences, from the corners of which the goldenrod still flung its banners to the breeze, though September and, mayhap, frosts had come. Beyond the fences were knolly pastures, cropped close except where the mullein, the thistle and the immortelle vaunted their immunity from the attacks of grazing herds; and still beyond were upland meadows, green with second-crop clover; and crowning all were forests beginning to glow with autumnal hues. Forth into such roads, pastures, meadows and woodlands were we wont to fare of a sunny morning, to hunt bees with our father, whose woodcraft was not the empty accomplishment of the man of this generation, but was attained on those same hills of western New York when he was a pioneer boy, and the deer and the wolves roamed those forests, and the beavers built their dams in the valley.

Our equipment for hunting was a bottle of diluted honey; a box with a sliding glass cover, containing pieces of empty comb; and a sharp stake, four feet high, topped with a cross-piece on which to set the box. When we were far enough afield, some unwary bee was lifted from its goldenrod revel and imprisoned in the box, where one of the empty combs had been filled from the bottle for her special delectation. Like a worthy bee, she began to fill at once; meanwhile, the stake was pushed into the ground and the box placed upon it, the cover removed, and we all retired for a little distance to watch. When the bee finally lifted herself and our honey into the air, we gave her closest attention. To make sure of the exact position of her bonanza, she always arose in a spiral, each circle being larger than the one before, and finally turned the spiral in a certain direction. When she suddenly darted away with almost the speed of a bullet, it was always the eyes of our father, blue as the sky against which the bee was outlined, that detected her direction; for young eyes, however keen, counted little against trained eyes in this competition.

Then there always followed a time of anxious waiting for the return of the bee. Meanwhile we stretched out on the dry sod in the sun and listened to the chirping of the crickets, or the sweet notes of the meadow larks and idly watched a hawk circle on even pinions above our heads; or we told stories of other days of successful bee-hunting. If the bee returned within fifteen minutes, all was well and we were confident that the tree was distant not much more than a mile. But if we had to wait a half hour we usually caught other bees and started over again, hoping to find some nearer colony. If our first visitor came back soon, and especially if she was followed by her anxious sisters, we were satisfied. With several bees flying in the same direction, it was always easy to get the "line," which we marked by some peculiar tree or other noticeable object in the landscape. When several of our visitors were eagerly filling themselves with honey, the cover was shoved over them and they were carried for a distance along the line and then liberated, and the line from this new location ascertained. Thus were they followed, up hill and down dale, and even through woodlands; finally, we would come to a place in a forest where we could follow the line no farther, and then we took our first lesson in geometry by getting a cross-line. This was done by carrying some of the bees in the box for some rods to the right or left; and when they were established there we knew that at the apex of the angle made by the two intersecting lines stood the bee-tree.

The triumph which filled us when we finally discovered that stream of black particles entering some knothole or the broken top of a tree, made us breathless; and all the way home we tried to temper our excitement so as to make the announcement of our discovery with a nonchalance characteristic of invariably successful hunters.

However, we were by no means always successful. Sometimes it would be too late in the day before we established a line; and again a line would lead us in a disgusting fashion to some unsuspected apiary; and now and then in a woodland tangle we crossed and crisscrossed lines, nearly breaking our necks gazing into tree-tops, all to no avail. Then, too, even our victory might be tempered by conditions. If the bee-tree were small, we judged it contained little honey; if the tree were valuable, we doubted if the owner would allow it to be cut. As a matter of fact, we seldom cut a bee-tree; and when we did, we wrested from it a combination of rotten wood, bee-bread, crushed brood and bees that made a potpourri which would prove disastrous to the enfeebled stomachs of this generation. But, though we rarely cut a bee-tree, bee-hunting lost none of its fascination. For what could be more delightful than long days spent in the autumn sunshine, enlivened by an occupation vitally interesting that needs must be lazily carried on! So we never gave it up until the October frosts had killed all the flowers, and the fumes of the honey-comb that we burned failed to entice an enterprising bee from her winter quarters to our box.

APPENDIX

BEE BOOKS

The bibliography of bee literature is extensive. Scientists of all nations have contributed the results of their investigations on bee anatomy and bee physiology, and have made bee literature, as a whole, most profound and technical reading. However, there are among these books many that were written for popular audiences, and that deal with the practical side of bee-keeping; of such we add a few titles of those best known and of special excellence.

BENTON, FRANK. "The Honey Bee." Mr. Benton, who is our national expert of the Department of Agriculture at Washington, always writes practically, and has carried on experiments with races of bees, which the private bee-keeper could hardly afford to make. The enterprising bee-keeper should keep in close touch with Mr. Benton's bulletins.

CHESHIRE, FRANK R. Two volumes. This is one of the finest works that has ever been published upon the honey-bee and bee-keeping. It is delightfully written and has many fine illustrations.

COOK, PROFESSOR ALBERT J. "The Bee-keeper's Guide or Manual of the Apiary." This is the most extensive of all the bee-keeping manuals written for American bee-keepers. It deals with all phases of the subject minutely, and new editions are published frequently enough to keep the book up-to-date. Professor Cook has a wide reputation as a most excellent teacher, and bee-keeping is one of the subjects which he taught for years in the Agricultural College of Michigan.

Cowen, T. W. "The Honey-bee, Its Natural History, Anatomy and Physiology."

Cowen, T. W. "Bee-Keeper's Guide Books." The first of these little volumes is a clear and excellent account of the anatomy of the bee. The second is a concise and helpful book on the methods of bee-keeping as practiced in England.

Huber, Franz. "Nouvelles Observations sur les Abeilles," published in 1792. English translation in 1841. This classic in bee literature is one of the most delightful of all the bee books that have been written. It shows the careful methods of this blind scientist who has given us more of the understanding of the bee and its life than any other investigator or writer.

Hutchinson, W. Z. "Advanced Bee Culture."

Hutchinson, W. Z. "Comb Honey." Mr. Hutchinson is one of our most successful bee-keepers, and he writes clearly and understandingly of his methods.

Langstroth, L. L. "The Hive and Honey-Bee." This classic in American apiculture has been revised and kept up-to-date by the scholarly Dadants', father and son, who are well known on two continents as successful bee-keepers. This book written by the father of American apiculture is comprehensive, and is good literature as well as good bee-keeping.

LUBBOCK, SIR JOHN. "Ants, Bees and Wasps." Although this is a book of scientific experiments, it should be read by every bee-keeper. No other book tells so well the patience and ingenuity necessary to discover what the bee knows and why it does certain things.

MAETERLINCK, MAURICE. "The Life of the Bee." This exquisite piece of literature and social philosophy has attracted much attention, and has introduced the world at large to the wonderful life of the honey-bee in such a poetic and dramatic manner, that most people have regarded it as a work of fiction. Maeterlinck is said to be a practical apiarist, and his book is based upon the facts of bee life as he understood them at the time the book was written. Though some of his facts be questioned, yet probably his statements are no more doubtful than would be those of almost any bee-keeper should he try to write what he thinks he knows about bees. Maeterlinck is the Homer of the bees and, therefore, he has a right to poetic license.

MILLER, DR. C. C. "Forty Years Among the Bees." This is a simply told history of the experiences of a successful bee man. It is a most honest and often a humorous record of bee-keepers' successes and failures.

MORLEY, MARGARET. "The Bee People."

MORLEY, MARGARET. "The Honey-Makers." The first of these books is written charmingly and simply for children and covers in an interesting manner the life of the bee. "The Honey-Makers" gives a most extended account of the relation of bees to men, giving extensive quotations from Hindu, Egyptian, Greek and Italian literatures and also a most interesting chapter on the curious superstitions and customs regarding the honey-bee.

QUINBY, MOSES. "Mysteries of Bee-Keeping," revised by L. C. Root and now called, "Quinby's New Bee Book." This is a simple, straightforward account of a practical man's dealings with bees.

ROOT, A. I. "ABC of Bee Culture." The author may be pardoned if she speaks with special enthusiasm of this book, as Mr. Root was the special teacher that helped the Comstock apiary achieve success. The interesting and truly human way that Mr. Root refers to bees is not only inspiring, but is also most practically helpful. The "ABC of Bee Culture" is arranged conveniently, encyclopedia fashion, so that the discussion of any subject in it may be readily found. Every page of it is interesting, and is based upon the actual experience of a man who is at once a keen observer, a sympathetic friend to the bees, and a most successful apiarist.

Printed in Great Britain
by Amazon